D0024978

SENSORY INHIBITION

THE

HERBERT SIDNEY LANGFELD

MEMORIAL LECTURES 1965

In late October and early November of 1965 there were delivered in Green Hall, in the Department of Psychology, Princeton University, the first of a series of lectures established to honor the memory of Herbert S. Langfeld. It is intended that the lectures, made possible through the devotion and generosity of Mrs. Langfeld, will recur annually and, in addition to presenting a summary of a major development of interest to the psychological public, will provide an opportunity for students and staff of the Department to benefit from close association and communication with the lecturer.

We consider ourselves fortunate that the series was inaugurated just at a time when Professor Georg von Békésy, the first Langfeld Memorial Lecturer, was organizing his thinking about the experiments on sensory inhibition reported in this volume. We are sure that Professor Langfeld would have agreed with us that the Békésy lectures constitute an important scientific and educational event.

The Herbert S. Langfeld Memorial Lecture Committee:

J. L. Kennedy · J. M. Notterman
C. C. Pratt · E. G. Wever
F. A. Geldard, Chairman

SENSORY INHIBITION

BY GEORG VON BÉKÉSY

PRINCETON, NEW JERSEY
PRINCETON UNIVERSITY PRESS
1967

L.C. Card: 66-17713
S.B.N. 691-08612-5
Printed in the United States of America by
Princeton University Press, Princeton, New Jersey
Second Printing, 1970

PREFACE

The experiments described in this lecture series have a long history; they began in 1927 and have continued to the present day. They suffered many interruptions, but they represented a sort of thread running through all my thinking for nearly forty years. The results never satisfied me, and yet they seemed almost of themselves to fall into a pattern. More and more it became clear that lateral inhibition is a common feature of all the sense organs and not a speciality of the retina, and that it is a simple consequence of the lateral interconnections of sensory nerve fibers. It was this generality that seemed to be important because it interconnected, at least in one respect, the fields of vision, hearing, skin sensations, taste, and smell. This interconnection made it possible to plan new experiments for one sense organ almost as a repetition of an experiment done with another. It was surprising in many instances to find even the numerical values to be closely similar.

These experiments were carried out in many institutes. They began in the Hungarian Institute for Research in Telegraphy and in the Physical Institute of the University in Budapest, then continued in the Royal Institute of Technology in Stockholm, at Harvard University, in a number of hospitals, and even in some private doctors' offices. I am deeply grateful to all these institutions with their facilities, staffs, and directors for having helped me to carry out the experiments.

I am very thankful further for the financial

support given me by several foundations. In Hungary it was the Széchény Foundation that supported my research for many years—and without ever asking for a progress report. In Sweden I was supported by the Karolinska Institute in Stockholm. In the United States the Office of Naval Research sponsored my research for many years. More recently it was the National Institutes of Health and the American Otological Society that made the continuation of this research possible.

Furthermore I am extremely thankful to a very large number of my subjects who were willing in many instances to work with me without any compensation even in situations where the results of the experiments were not promising. Some of these subjects were so kind as to repeat earlier observations many years after the first tests. These repetitions were especially valuable in consolidating results that were often difficult to obtain.

I am very thankful to the committee of the Langfeld Lectures and to its chairman, Professor Frank Geldard, who invited me to give the Herbert S. Langfeld Memorial Lectures. This assignment gave me a framework into which to place the phenomenon of lateral inhibition and to bring out the magnitude of inhibitory effects. The timing of this invitation was very fortunate, because I had just discovered how general is the similarity of inhibitory processes in the different sense organs. It was a very great pleasure to lecture at Princeton University, mainly because of the interesting discussions and the opportunity of demonstrating some of the experiments to the audience. In this relation I am grateful to Dr. Carl Sherrick for helping me to assemble the demonstrations.

One of the most difficult questions was whether the lectures should be published or not. My English is still extremely poor, but during a lecture I can use my hands in an effective way. Unfortunately in so-called idiomatic English rendered in print these ancillary aids are missing, and I needed a much more sophisticated type of assistance. It was Professor E. G. Wever who solved the problem by being willing to edit the lectures. As everyone knows, editing a poor English paper is much more difficult to do than translating a not-so-poor German paper into English. It is therefore E. G. Wever whom the reader and I have to thank that the lectures are being published. His assistance was especially valuable in calling my attention to portions of the lectures that were ambiguous in their phrasing. This criticism not only has made the lectures easier to read, but has shortened them without loss of content.

Many of the figures were published in my earlier publications and others are taken from the articles of colleagues. I here express my gratefulness to the authors and to the publishers of the journals who have kindly granted permission for this reproduction. The author and journal are named in the caption of each figure, and the complete citation appears in the list of references. The numbers in brackets refer to my own publications, to be found in the Author's Bibliography. The publishers to whom this acknowledgment is made are the following:

Acta Oto-laryngologica
American Association for the Advancement of Science (*Science*)

American Physiological Society (*Handbook of Physiology* and *Journal of Applied Physiology*)
Cambridge University Press (*Journal of Physiology*)
The Company of Biologists, Ltd. (*Journal of Experimental Biology*)
Elsevier Publishing Company (*Le Prix Nobel 1961* and *Proceedings of the Third International Congress on Acoustics*)
S. Hirzel Verlag (*Akustische Zeitschrift* and *Physikalische Zeitschrift*)
The Johns Hopkins Press (*Bulletin of the Johns Hopkins Hospital*)
Holden-Day, Inc.
National Academy of Sciences (*Proceedings*)
Optical Society of America (*Journal*)
Rockefeller University Press (*Journal of General Physiology*)
W. B. Saunders Company
Springer-Verlag (*Elektrische Nachrichten-technik, Handbuch der normalen und pathologischen Physiologie, Naturwissenschaften*, and *Zeitschrift für Hals- Nasen- und Ohrenheilkunde*)
U.S. Army Medical Research Laboratory, Fort Knox, Kentucky (Report 424)
University of Chicago Press
John Wiley and Sons
Wistar Institute of Anatomy and Biology (*Anatomical Record* and *Journal of Cellular and Comparative Physiology*)

July 1966

GEORG VON BÉKÉSY

Laboratory of Sensory Science
University of Hawaii
Honolulu, Hawaii

CONTENTS

SENSORY INHIBITION

Willst du ins Unendliche schreiten,
geh nur im Endlichen nach allen Seiten.

GOETHE

I · ADAPTATION AND INHIBITION AS A MEANS OF SUPPRESSING AN EXCESS OF INFORMATION

There are many ways in which a physical activity may be so modified as to cause a human observer's perception of it to depart widely from the indications of physical apparatus. We are all aware of the existence of a sensory threshold, adaptation processes, and nonlinearity, all of which represent departures of perception from the regular variations of magnitude in a physical stimulus. A further process is inhibition, whose effects are equally unexpected and often even more profound than these others.

My first concern with the subject of inhibition goes back to an article in the *Physikalische Zeitschrift* in 1928, in which an appeal was made to the interactions of nerve fibers supplying adjacent regions of the basilar membrane in order to explain the sharpness of pitch discrimination in the presence of a high degree of damping. A theory of tonal contrast was suggested along the lines of Mach's law of brightness contrast as applied to adjoining visual fields. At the time my attempt to introduce psychological processes into an area that was then held to be purely mechanical met with great opposition on the part of my associates in the Physics Department. I could appreciate the point made by my critics in this connection, that physical measurements excel all others in precision and ought to be preferred in any problem dealing with vibratory actions. And yet I continued to be concerned with the differences between perceptual observations and physical measurements. My respect for psychological observation was further enhanced when I learned that in some situations, as in the detection of weak stimuli, the sense organs often exhibit greater sensitivity than can be demonstrated by any purely physical procedure. Therefore my interest in the psychological processes was unquenched, and my concern with inhibition continued through the following years.

Fig. 1. Ohm's law was discovered in two steps. The early form is shown at *a*, and the revised form at *b*.

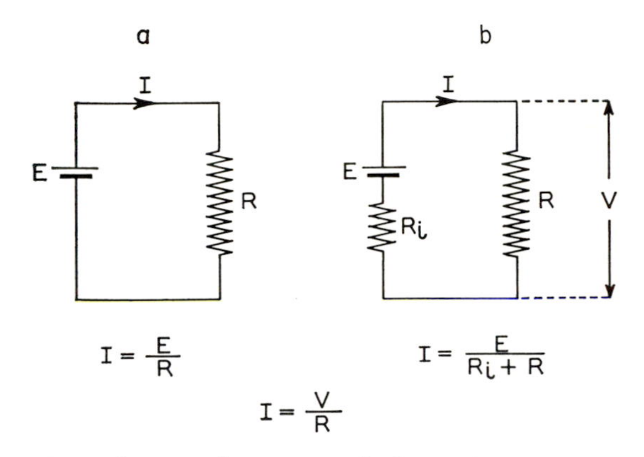

It is obvious that in psychological experiments we cannot isolate the system of interest and control all the variables. A single uncontrolled variable can make all the observations imprecise. The same is true of experiments in physics also. When Ohm first described his famous law of electricity the current was calculated according to formula *a* of Fig. 1 as determined by the voltage of the battery and the resistance, $I = E/R$. This result was not at all precise. It was probably Poggendorf, the editor of the *Annalen der Physik*, the journal in which Ohm's formulation first appeared, who realized that the battery must itself have a resistance along with the external resistance. The introduction of the internal resistance R_i, as indicated by formula *b* of this figure, made Ohm's law one of the most precise in the field of physics. There is hope that in the course of time psychological observations also will disclose some of the variables still unknown. In the discussions to follow, however, it must be borne in mind that many of the variables have not been controlled, or have been controlled only in part, and therefore the results cannot be as exact as in an ideal experiment.

A further limitation in this series of discussions needs to be mentioned. I shall make no attempt to treat the subject of inhibition in a complete way, and many areas of considerable interest and importance will not be mentioned at all. Instead, I shall follow (in only a partly systematic way) the course that my own thinking and experimentation has taken in dealing with this subject, and show how the subject has continually intruded itself in my investigation of sensory systems.

In a sense, this way of approach to our subject may be called the mosaic method: a good many observations of inhibitory phenomena and effects are made, one after another, and placed in a pattern that the results themselves seem to require, but without a preliminary conception of what the pattern ought to be. I can compare this procedure with one that I once observed in the work of an artist in Paris. The artist was painting a picture of an old building near the medical school. He spent a long time mixing a color on his palette that would exactly match a spot on one corner of the building, and when he was satisfied with his mixture he put a small spot on his canvas at the proper place. He then went on to another small spot, and at the end of the first day he had only four or five spots. After a week the spots numbered about a hundred, but they matched precisely the colors of the different small areas of the building, a road, and some trees. The artist showed me another painting done in the same way, but in which at the end the dots of color had been smeared together to give a statistical mean. Unfortunately this painting lacked the realism and freshness of the unfinished one, and for me it had lost its attractiveness.

My research plan for the subject of inhibition and this discussion of it will be left as a "dot" painting, without statistical treatment. This method has many advantages. Each "dot" or individual aspect of the treatment can be considered separately, and rests on its own merits. Criticism of a part does not imply an impeachment of the whole, and an individual error can be rectified without modifying the entire structure.

The problems of inhibition are involved in basic discussions of nerve action. One of these discussions, representing an enduring controversy, has to do with the localization of neural activity in the brain. At one time it was held generally that localization in the brain is not specific but rather that different areas may be involved from time to time in a given kind of activity. The present trend is away from such "mass action" and toward a high degree of specificity. An important step was taken by Hoffmann in 1914, when he showed in the crayfish that peripheral inhibitions depend upon specific nerve fibers. Later Marmont and Wiersma (1938) found that certain nerve fibers have no function but inhibition. These observations were further substantiated by the work of Renshaw (1941) and Eccles (1957).

This development shows clearly that as experimental techniques improved the assumption of specific inhibitory nerve fibers became more and more firmly established. In my opinion this parallel between the improvement of technique and the reliability of the observations is proof that we are on the right track. This simple matter of localization indicates that taking inhibition into consideration as a part of investigation in this field might change the direction of the research.

Adaptation and inhibition in information reduction

A few years ago the transmission of information between distant points was one of the most desired goals. The first long-distance telephone conversation between two countries of Europe, and even more dramatically the communication between two continents, was considered a triumph of applied science. Everyone looked forward to the time when pocket radios would permit two persons anywhere to communicate with one another. The requirements were simply a high degree of sensitivity and slight distortion.

The success during this period justified what I should like to call the "input-output complex," represented in Fig. 2. Here we have a black box with an input on one side and an output on the other. The black box can mean anything: a mechanical system, an electrical network, or the brain. Whenever an input is introduced there is an output on the other side. For example, a spot of light on the retina will produce an electrical discharge, which passes through the electrical network of the brain to produce finally a sensation at the cerebral cortex. In a way this is an electrical formulation of the causality principle.

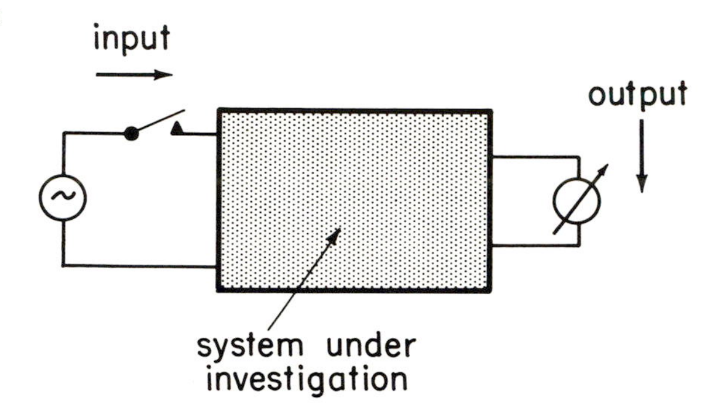

Fig. 2. The input-output system, with a generator on the left, a receiver on the right, and a "black box" in the middle, is the simplest electrical representation of the principle of causality. From U.S. Army Medical Research Laboratory, Fort Knox, Kentucky, Report No. 424 [93].

It was assumed that there was no limit to the amplification of light intensities, light input, and electrical potentials. The weakest signal ought to produce a practical output. A severe disappointment came with the discovery that any amplifier generates noise and that the noise limits the usefulness of a transmission system. The important thing in a communication network is not the output level attained but the signal-to-noise ratio, for it is this ratio that determines our ability to recognize the signal as distinct from the noise. In the nervous system also it was found that "noise" is always present, in the form of a general background of spontaneous activity, and a sensory effect has to be identified in the presence of this background (Hoagland, 1932). This problem is still with us.

For me the euphoria of the input-output concept lasted only a few years. I shall never forget the time when as a young communications engineer I could first talk over a line from Budapest to London. It seemed that henceforth it would be possible to answer, in the field of communication at least, the questions that a government asks of a communications engineer. Naturally we tried to make more and more improvements in our transmission systems. In a few years we had a regular wire communication system and a long-range radio system. Then later we had a short-wave system and finally an ultrashort-wave system. This development produced an unexpected problem: which of these four systems is the best? I had the impression that it was easier to develop a new transmission system than to decide which one of several systems was the most suitable. The evaluation problem caused us to stop our development of systems and to become involved in questions of detail. We had to measure the distortion of the systems, their signal-to-noise ratios, their stability over long periods, ease of operation, economy of construction and maintenance, and so on. We tried to develop "automats" that would always select the best transmission line, but it soon became clear that in selecting one we had to eliminate the others. In a manner of speaking we "inhibited" all but one among four or five possibilities. Few engineers were willing to take this responsibility. It was in this situation that I first became aware of the fact that not only in communications engineering but also in other fields like physiology the reception of information is not a simple matter, and it is necessary to reduce unwieldy complexity by cancelling out some of the information and possibilities of its treatment.

At a recent meeting of the Physical Society, where I gave a talk, there were 730 contributed papers and 50 invited papers, and in each of these there were about 5 figures, making altogether about 3000 figures, all confronting the eyes of the members within a period of three days. It is said that today we have an age of explosions: the atomic explosion, the population explosion, and worst of all the information explosion. Survival requires that we discard the unimportant portions of this information.

A decade ago there was a development of communications and information theory. Already we have too much of this theory, and it is necessary to develop an inhibition theory. We shall need to discover a way of measuring the loss of information caused by a given amount of inhibition. A possible measure is the number of bits lost when the information is passed through a system divided by the number of bits introduced at the input. Such a measure I consider more important in physiology and psychology than in communications engineering.

We know that any information that we have at hand contains a number of small disturbances. These are usually eliminated by a statistical treatment. The mean value of a series of measures represents the inhibition of many small and unwanted bits of information, but this procedure does not go far enough. What we are interested in is a direct form of inhibition that cancels out a whole group of unwanted information. The problem is of far greater scope than statisticians have ever dreamed of.

The threshold

The simplest way to get rid of information is to reduce the sensitivity of the receptors. This method is used effectively in all complex living systems. For example, we know that the organ of Corti of the ear is sensitive to displacement. This sensitivity is so great that one can almost hear the Brownian movement of molecules. Yet at the same time the organ of Corti, as living tissue, requires a constant blood supply, which is provided by the streaming of blood cells through capillaries. This streaming of the blood cells involves a completely irregular series of impacts with the capillary walls. Also the blood stream is not a steady flow; because the heart acts like a pump, there are periodic pressure variations extending even to the capillaries. These pressure variations greatly exceed the magnitudes of the sound pressures that we wish to detect with the ear. Hence we should expect to hear our own heartbeat with tremendous loudness. We do not do so because, for one thing, there is a factor of frequency differentiating between external sound stimuli and the sounds of the heartbeat. The circulatory pulsations are of low frequency, and the solution that nature made to the problem was to reduce the sensitivity of the ear to the lower frequencies while leaving the sensitivity unimpaired in the range between 1000 and 4000 cycles per second (cps). Thus it is not surprising to find that the threshold sensitivity for frequencies around 20 cps is nearly 10,000 times less than for frequencies around 1000 cps. These relations may be seen in Fig. 3.

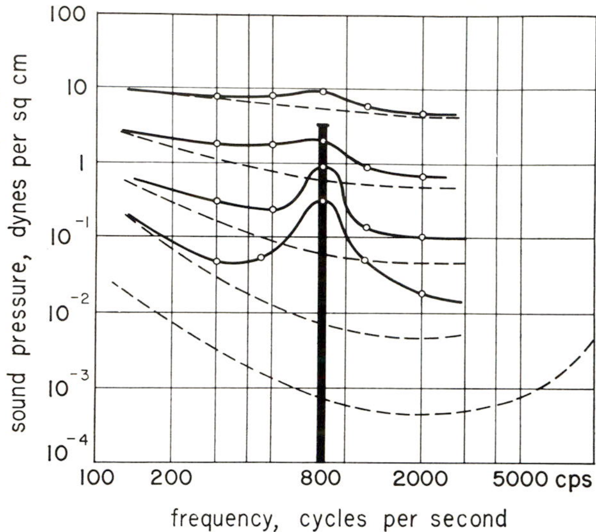

Fig. 3. Equal-loudness curves for the human ear, in the middle range (broken lines). After exposure to a strong tone of 800 cps, these curves are modified (solid lines). From *Physikalische Zeitschrift* [4], modified.

In vision we have a similar situation. The eye also requires an abundant blood supply. As is well known, any body with a temperature above absolute zero produces radiation. The higher the temperature of the body, the shorter are the wavelengths of heat or light emitted by it. Most people know from experience that as we begin to heat a piece of metal it first glows with a deep red color, then yellowish red, then white, and then bluish white. For the very high temperatures obtained with sparks the color is blue, and, for even higher temperatures obtained by electron bombardment, the emitted rays become so short as to be invisible and are called X-rays.

Our own bodies have a temperature of 37° C (310° K) and at this temperature there is infrared radiation. It is necessary for the eye's sensitivity to be sharply reduced in this region of the spectrum to prevent our constant seeing of red light. I would say that nature did a marvelous job in varying the sensitivity as a function of wavelength so that with closed eyes in a dark room we see only a trace of redness.

In other sense organs like taste and smell it is likely that the internal stimuli are not as constant in physical characteristics as those in hearing and vision, and therefore nature used the process of adaptation to eliminate the background stimulation. Adaptation is mainly responsible for the fact that (usually) we do not taste our own saliva. The olfactory apparatus uses a different device: there is a construction that determines aerodynamically that outside stimuli are favored over inner ones. The olfactory epithelium is in a blind passage so placed as to receive eddy currents during inhaling to a greater extent than during exhaling.

The loss of information through masking is a well-known effect. We are all aware of the fact that noise in a room can make it impossible to understand speech. Nature does not use this means of reducing information because it is unreliable and difficult to control; a masking sound is as likely to cancel the wanted information as what we would dispense with. Therefore nature has used inhibition in preference.

The masking effect of noise was what led communications engineers to introduce the concept of signal-to-noise ratio. In sensory systems the role of inhibition is mainly to improve the signal-to-noise ratio.

Adaptation as a form of inhibition

Adaptation is a common process in living systems for the reduction of the effect of a stimulus. It is seen mainly as a progressive loss of sensitivity during a period of stimulation. It is easy to measure in hearing.

In Fig. 3 the lowest broken line represents the pressure amplitudes of various tones at the threshold. The other broken lines are the so-called "equal-loudness curves" at four different sensation levels (Kingsbury, 1927). These curves were obtained by increasing the sound pressure of a reference tone of 800 cps by 20, 40, 60, and 80 db above its threshold value and then matching the loudness of this tone at each level with tones of different frequencies. We note that as the loudness is increased the equal-loudness curves become more and more flat.

If now we present to one ear a preliminary tone with a sound pressure of 3 dynes per sq cm and determine the equal-loudness curves immediately afterward, we obtain a set of functions different from the former ones. These curves are shown in Fig. 3 by the solid lines. To determine their positions on the ordinate we have to make a comparison at each frequency between the loudness now obtained and the loudness perceived in the other ear, which has not been exposed to the "adapting" tone. For the lowermost of the solid curves of Fig. 3 the 800-cps tone had its threshold elevated (i.e., impaired) by about 50 db, and other tones suffered adaptations of lesser amounts.

The amount of auditory adaptation depends upon the loudness of the stimulus. For sound pressures near the threshold the adaptation is negligible. This fact can be demonstrated by recording continuously with a semi-automatic audiometer the threshold of hearing for a given constant frequency. To do so requires an equipment that gives an increase in sound pressure when the observer presses a button. When the observer hears the tone he releases the button, and the sound pressure begins to fall continuously. If he then finds that the tone is not heard he presses the button again, whereupon the sound pressure rises once more. The ups and downs of sound pressure are recorded on a graph as shown in Fig. 4. The lower tips are the values for which the observer is just unable to recognize the tone, and the upper tips are the ones at which he just perceives the sound. The middle between these two extremes may be called the threshold.

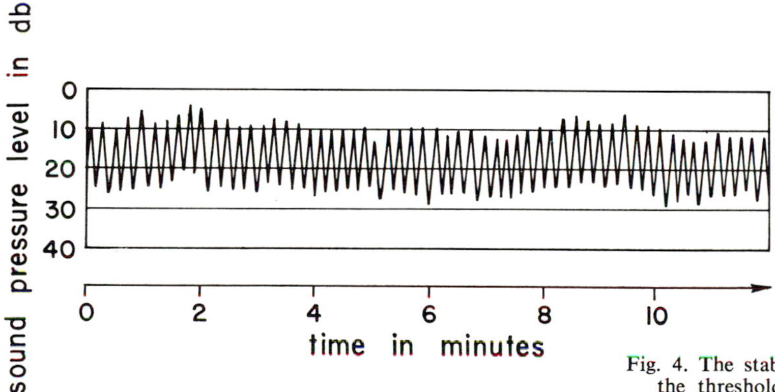

Fig. 4. The stability of the threshold for a 1000-cps tone for a human ear, obtained with a semi-automatic audiometer. Adapted from *Acta Oto-laryngologica* [49].

As may be seen in the figure, it is possible to record threshold values in this manner for 10 minutes or more without much change. It is clear that for the ear the sensitivity is not affected by presenting stimuli at the threshold level. For other sense organs, however, this is not the case.

The sensation of taste is one in which, as is well known, the magnitude of sensation diminishes greatly with time. Unfortunately it is difficult and cumbersome to present well-defined stimuli to the tongue. The concentration of the test solution must be changed immediately and continuously according to the rotation of a knob. There are several ways of constructing a mixing apparatus to do this, and one way is illustrated in Fig. 5. The apparatus consists of two peristaltic pumps whose speed of rotation is adjusted with a differential gear in such a way that there is a constant flow out of the

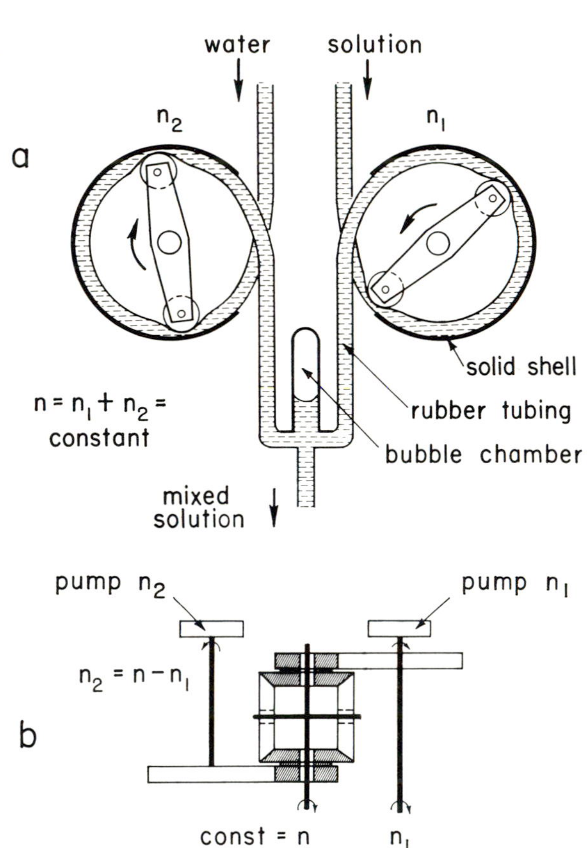

Fig. 5. Equipment for producing the stimuli used in experiments on taste. At *a* is shown the peristaltic pumps, and at *b* the differential mechanism through which the rate of flow of the final mixture was kept constant. From *Journal of General Physiology* [116].

mixer. The differential gear can operate so as to increase the quantity of solution and correspondingly decrease the amount of water to be mixed with it, giving a stronger test solution, or to decrease the solution and increase the water to give a weaker mixture. The complete setup is shown in Fig. 6, where the changes in one of the pumps are recorded to show the variations in concentration of the test solution over a period of time.

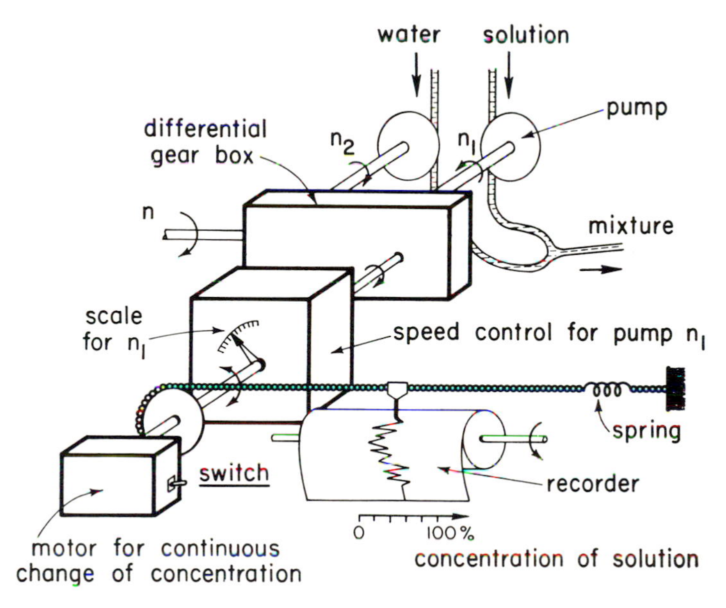

Fig. 6. Semi-automatic equipment for the recording of taste thresholds. From *Journal of General Physiology* [116].

To present the test solution to the surface of the tongue a plastic block as in Fig. 7 is used, with connections for an inlet and an outlet and a channel opening at the surface where the tongue is placed. (This block as shown is provided with two test loops, but only one is used for simple threshold determinations.)

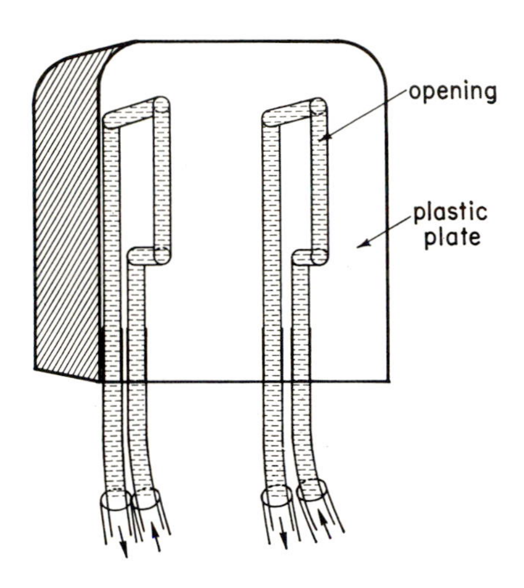

Fig. 7. Plastic plate with two openings, for stimulation of the tongue. From *Journal of General Physiology* [116].

As was done in hearing, this method is used to find the points at which a taste is just recognized and just not recognized. But the effects through time are very different from the auditory situation (Fig. 8). There is rapid adaptation, so that ever-increasing concentrations of a taste substance (hydrochloric acid in this example) are necessary for threshold sensitivity. In about a minute the sensation disappears: the available concentration will no longer stimulate the taste endings. Part *E* of this figure shows what happens when the observer moves his tongue laterally with respect to the window in the plastic block. New taste endings are brought into play and the sensitivity rises abruptly, and then declines as before.

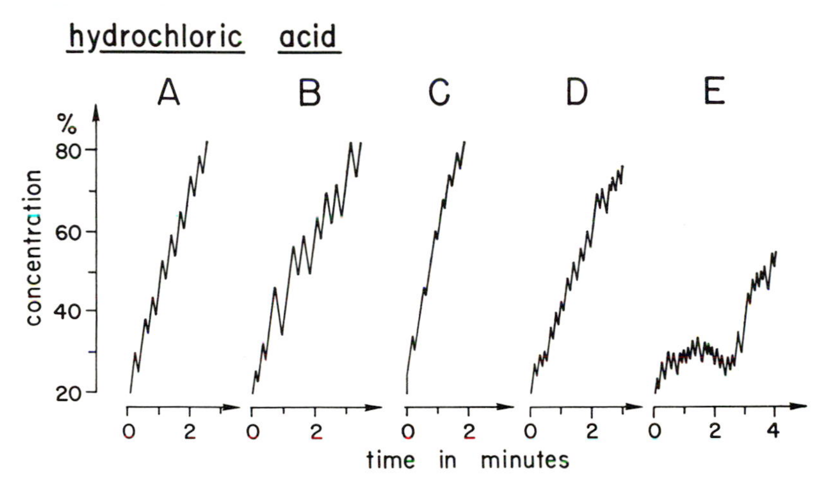

Fig. 8. Effects of adaptation on the threshold for hydrochloric acid (sour taste). Curves *A* to *D* represent different observers. Curve *E* is for an observer who moved his tongue during the recording. Note that a rise in the concentration needed represents a decrease in sensitivity. From *Journal of General Physiology* [116].

Adaptation in vision is almost of the same order of magnitude as in taste. The retina seems to lose its "pattern recognition ability" quickly if the pattern is maintained on one portion of its surface. As has been proved by Riggs and others (1953) an image that is made stationary on the retina disappears partly or completely in a few seconds. To carry out this experiment it was necessary to use a rather elaborate optical arrangement that would maintain the image at one place on the retina in spite of movements of the eyeball.

It is possible to demonstrate this phenomenon in a simpler way. If we illuminate a homogeneously painted wall surface in a dark room with a diffuse light we see some hazy dots or threads as shown in Fig. 9 floating in front of the equally illuminated surface. If we try to look closely at one of these dots we see it float away. The farther to one side the dot is seen, the higher is its speed of movement toward the edge.

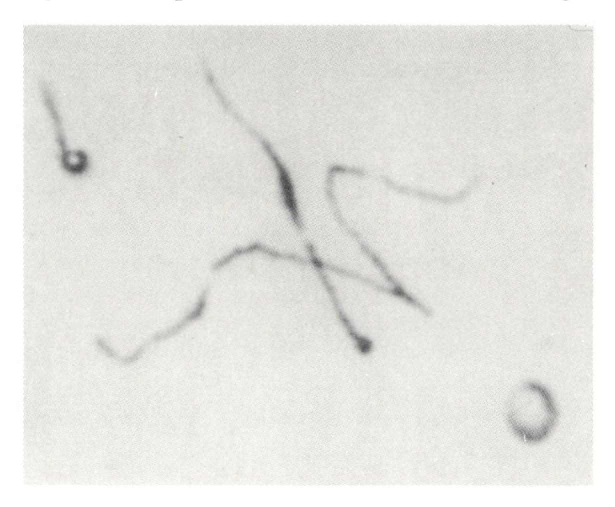

Fig. 9. Floating threads, seen by projection of the visual field on a homogeneous surface.

With a little practice it is possible to observe one of the fuzzy dots and to hold the eyeball still. Then it is surprising to find that the floating dot disappears in 1 to 1½ seconds. It returns immediately, however, if the eyeball is moved even slightly. These observations indicate that an image stimulating only a limited group of retinal cells remains in view for only a short time. Under ordinary conditions the continuous movements of the eye prevent loss of our visual sensations through adaptation.

That the transients of a stimulus are of utmost importance in vision was shown in an objective way in experiments on the eye of the horseshoe crab, *Limulus*, by Ratliff, Hartline, and Miller (1963). Even a moderate change in the stimulus intensity, such as doubling or halving, produced an immediate change in the discharge rate of a single unit of the optic nerve, as seen in Fig. 10. This figure shows the effect first of a doubling of the light stimulus, and then a return of the stimulus to its former level. The response frequency increases rapidly to more than twice its original value when the light is increased, and in about 0.5 sec returns to the base line. Then when the light intensity is dropped to its initial level, the frequency falls abruptly and soon rises to the base level once more.

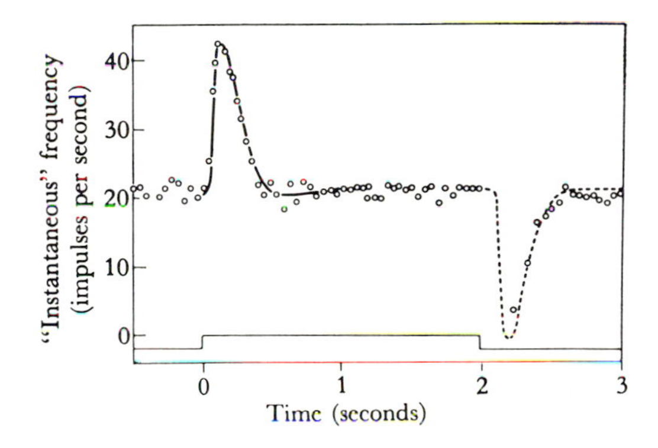

Fig. 10. Effects of increasing or decreasing the light intensity on the *Limulus* eye, as recorded from a single ommatidium. From Ratliff, Hartline, and Miller (*Journal of the Optical Society of America*, 1963, 53, 110-120).

The process of compensation

The processes of masking and adaptation reduce information by reducing receptor sensitivity, and this reduction continues even after the disturbing stimulus has ceased. In some situations, as when defense reactions are needed, this impairment of sensitivity can be a handicap. A more suitable method of eliminating unwanted information is through the process of compensation.

One of the best-known compensation systems is found in the vestibular organ. The statoliths of both sides continuously relay neural activity to the higher center. If the head is in a vertical position, the neural discharges from the two labyrinths to the center are equal. They compensate for one another, and no activity is transmitted to a higher level. When the head is tilted to one side, there will be an increase of activity in one labyrinth and a decrease in the other. The compensation is destroyed, and activity will be transmitted to the higher center indicating the direction of tilting of the head. This mechanism avoids an overloading of the higher center with information that is unimportant, such as the fact that the head is in a normal position.

Compensation methods are widely used in electrical and mechanical engineering, mainly because during the equilibrium position, when the communication lines are not being used, there is no expenditure of energy. Indeed, to the physicist the compensation methods seem the most efficient, and should be used more generally in organisms to avoid a needless waste of energy.

An interesting application of the compensation idea was made by Holst and Mittelstaedt (1950) in an explanation of bodily movements. For example (see Fig. 11), if the hand is moved, it is assumed that the central nervous system Z transmits to a lower level Z_1 an advanced program as to how the movements should be performed. This program is relayed to the nerve fibers and muscles and executed by them. As the movements occur the proprioceptors relay various bits of information back to the level Z_1.

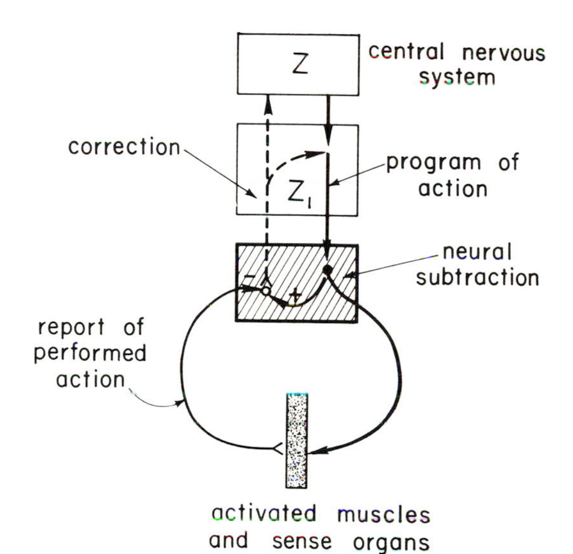

Fig. 11. Schema showing the compensation principle in neural action. From Holst and Mittelstaedt (*Naturwissenschaften*, 1950, 37, 464-476).

If the performance and program are identical then no further action is taken at level Z_1: the whole motor process is compensatory. If, however, there is a discrepancy between performance and program, the difference will be transmitted to the higher center Z, where the programming will be revised and corrected. This is a simple scheme for reducing the load on the higher center for routine, often-repeated actions.

It is obvious that a system of compensation of this kind requires some unique equipment. It requires a programming device, a receptor system in the muscles to indicate to what degree the program was carried out, and a compensating unit to match the outgoing and incoming information. Everyone will agree that at the present time we have no idea how these things are achieved. We do not have the knowledge even to outline clearly the requirements of such a system. Yet the model has great value in suggesting the types of neural processes that we should look for. Holst and Mittelstaedt believed that compensation systems account for much of the performance of the simple animals.

Before we end the account of the methods used by nature to get rid of too much information without resorting to inhibition, it is necessary to mention the simple loss of memory. We have two types of memory: short-acting and long-lasting types. It is the short-acting type of memory that serves immediate needs and then relieves the nervous system of further retention of the information. The long-lasting memory processes thus are protected from overloading.

Most of the inhibitions involved in getting rid of information depend in large degree upon conditions such as health, temperature, and fatigue. Perhaps some of the most dramatic changes occur from alterations of the blood supply. Nothing is more disturbing than the complete lack of short-time memory that may be seen in a person suffering senility or arteriosclerosis of the brain, which results from an inadequate blood supply.

Early history of inhibition

Inhibition must have been known in connection with motor nerves centuries ago, after the discovery of the two antagonistic muscles for movement of arms and legs. It became obvious that to move the arm, one muscle must contract and the antagonist must at the same time relax, for the muscles act only by shortening. This observation led to the conception of inhibition of one muscle during the activation of its antagonist.

As early as 1662, in his work *De Homine*, Descartes described clearly the action of two muscles in producing lateral movements of the eye, and their interconnection so that when one muscle was contracted the other was relaxed. Figure 12 reproduces the original drawing that shows how one muscle swells and the other shrinks. This idea was elaborated further in Descartes' later writings.

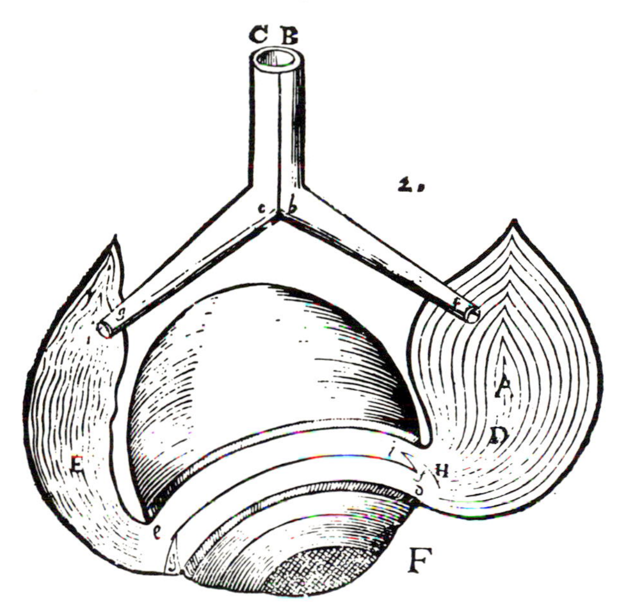

Fig. 12. Reciprocal innervation of the eye muscle, according to Descartes, 1677.

The general importance of inhibition was recognized by the two brothers Weber in 1845, when they discovered that stimulation of the vagus nerve produces a slowing or stopping of the heart. This was an important discovery because it revealed that inhibition occurs through the operation of well-defined neural pathways.

Another form of inhibition was shown by Sechenov in his *Reflexes of the Brain* in 1863.

He found that certain reflex movements of the frog can be influenced by exciting the midbrain with electric currents or by applying a salt crystal. This is central inhibition. Later developments in this field were carried out by Sherrington, and described in his well-known book on the integrative action of the nervous system.

Recent progress in the field of inhibition was summarized in the proceedings of the Friday Harbor Symposium on Nervous Inhibition edited by Ernst Florey (1961).

Lateral inhibition in the sense organs was first described, as far as I know, by Ernst Mach (1866, 1868) in his description of what we call the Mach bands. This phenomenon was quite neglected for nearly 80 years, because it was regarded as a unique effect and was not interpreted as a result of inhibition. The reason may be that lateral sensory inhibition is somewhat different from the inhibition of motor activity. It was largely the work of Hartline (1949) and of Ratliff, Hartline, and Miller (1959, 1963) on the eye of *Limulus* that stimulated a fresh interest in sensory inhibition.

Today our questions concerning inhibition are on a very general and extensive level. Thus as our electrophysiological methods improve we are discovering interconnections between more and more of the parts of the nervous system, so that we must conceive of this system as a vast meshwork of interconnected units. For an electrical engineer this picture immediately presents the problem of how such a system can operate with so much feedback without going as a whole into oscillation. In the early experience of building amplifiers, this is just what usually happened. To avoid such oscillations, the nervous system must somehow differ from the usual electrical networks used in engineering. At the moment it is not possible to say what the differences are, but it is worthy of note that something much like a state of oscillation occurs in tetanus, Parkinson's disease, and strychnine poisoning.

A second problem arises from the fact that in most nervous systems there is constant activity in the absence of a stimulus. This spontaneous activity seems to be similar to the activity produced by a stimulus. In hearing, we have a spontaneous activity in the auditory tract that a person seems not to hear. It would be interesting to determine where and how this activity is eliminated so that it does not produce a disturbance.

Inhibition is not a minor phenomenon

It is often assumed that inhibition is a minor side effect, modifying a stimulation pattern only in small degree. Thus it is supposed that special instrumentation is needed to observe sensory inhibitory effects. Only in the last few years has it become clear that this assumption, which arises out of the input-output concept, is not at all valid.

Perhaps the easiest way to show that sensory inhibition is a highly important phenomenon is to study the patterns of sensation on the surface of the skin produced by well-defined stimuli. If a vibrating tip is placed on the surface of the upper arm as in Fig. 13, then traveling waves are set up along the skin surface much like the waves that may be seen on the surface of water (Keidel, 1952). These traveling waves move away from the vibrating tip, as may be proved by the use of stroboscopic illumination.

Fig. 13. Traveling waves on the skin, at two different frequencies. From *Journal of the Acoustical Society of America* [74].

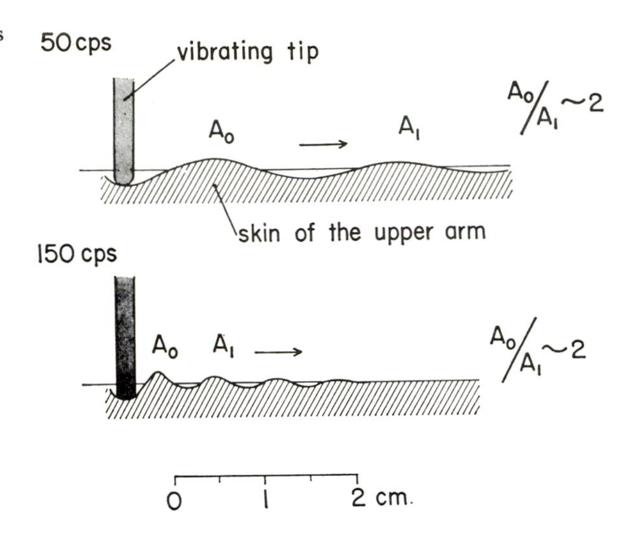

For vibrations of 50 cps, the wavelength of the skin waves on the arm is of the order of 2 cm. It is surprising to see that these waves are but slightly damped; in general their amplitude is reduced by about half for every wavelength of travel.

The skin is a complicated medium, mainly because it is a combination of a stiff surface layer and a soft under tissue. Therefore it has high dispersion for mechanical vibration, and the speed of the traveling waves depends on the vibration frequency. As Fig. 13 shows, an increase in vibratory frequency from 50 to 150 cps reduces the wavelength from 2 to 0.6 cm. In this range, however, the damping is nearly constant.

At 50 cps under good stroboscopic illumination we can see the traveling waves running along the whole arm and even to the surface of the chest. The deformations along the skin surface can be seen easily, and we would expect that the vibrations would be perceived over a large part of this surface. The surprising fact is that they are felt only over a small area in the region of the vibrator, an area lying within about 1 cm or less of the vibrator tip. The movements beyond about half a wavelength from the vibrator produce no sensation: I shall say that the activity away from the point of application of the vibrator has been inhibited.

This type of inhibition can be demonstrated in many ordinary situations. When a person speaks he produces vibrations of the vocal cords that are very extensive, and can be detected along the entire surface of the chest by applying an ordinary phonograph pickup. In general, we are not aware of these vibrations except during the formation of special sounds like the vowels "a," "o," or "u."

An even more interesting experience is the following. If a person stands on a vibrating platform, the vibrations transmitted to the soles of the feet are propagated along the entire body (Fig. 14). These vibrations are conducted largely by the bones and ligaments, and for low frequencies are subject to only slight damping, as may be observed by illuminating the skin with stroboscopic light. Yet if we ask the person how far up on his body he can feel the vibrations, he will report for low frequencies that this distance is much less than his entire height. For 150 cps at an amplitude for which we can stroboscopically observe the vibrations as far as the hip, the person will say definitely that the sensations extend only two or three inches. Again we find that in large part the effects pro-

Fig. 14. Propagation of vibrations along the body, on stimulation of the hand (shown on the left) and the foot (shown on the right). The horizontal width of the outlines represents the amplitude of vibration, on a logarithmic scale. From *Akustische Zeitschrift* [33].

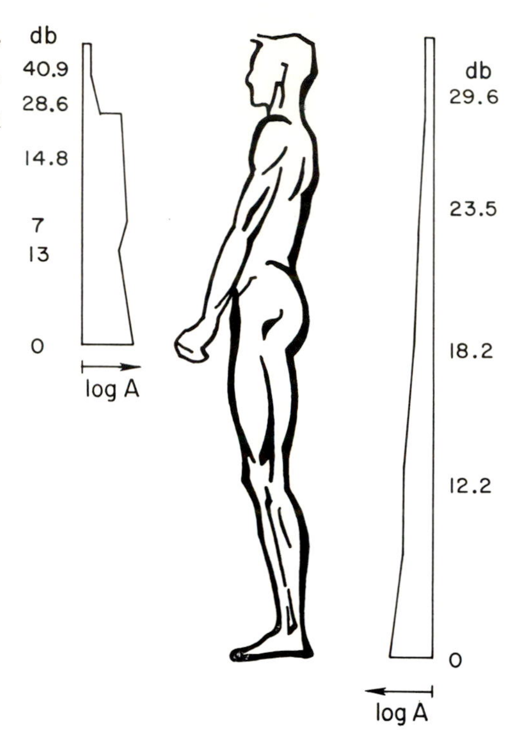

duced in joints, muscles, and skin have been inhibited.

Experiments with the blind spot of the retina

The phenomena of sensory inhibition are readily demonstrated in vision, where they play a major role in the production of a sharp image. As is well known, the lens of the eye departs widely from the glass lenses used in optical equipment. The optical lens is ground with great precision from a homogeneous piece of glass carefully chosen to exclude bubbles or flaws. The lens of the eye is very different. As Fig. 15 shows, it consists of a series of concentric layers like

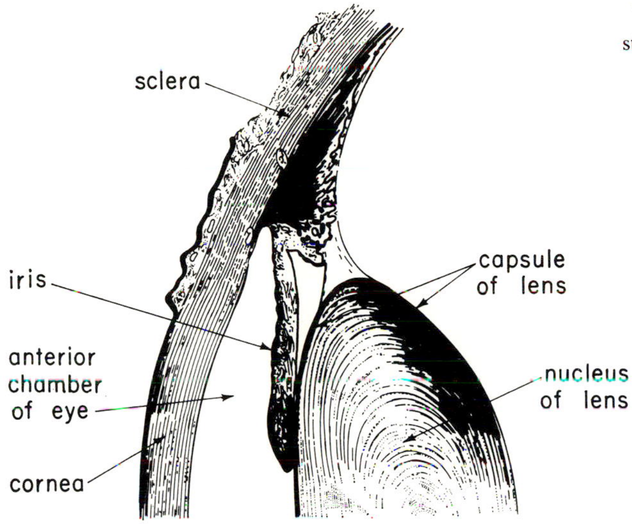

Fig. 15. The lens of the eye, and accessory structures, seen in cross section. After Gray.

the layers of an onion. We may regard it as a set of lenses enclosing a central one. Such lenses have advantages and disadvantages, and have some analogous relations to the "lens" of an electron microscope.

The onion-layer lens produces a sharp image, but around the image there is a certain amount of "fuzz." We can see this effect if we look at the image of a point of light cast on the back surface of the excised eye of a cow. Physically there is a dispersion of light outside the image of the object, but ordinarily we do not see the dispersion in our own eye.

We can set up an experimental situation in which this dispersion becomes visible. Figure 16 shows how a spot of light A_1 is projected upon the retina at a point to form the image A_2. Such an image seems sharp, but as the upper sketch of Fig. 17 shows there is in actuality a small

Fig. 16. Optical dispersion in the eye, and the blind-spot experiment.

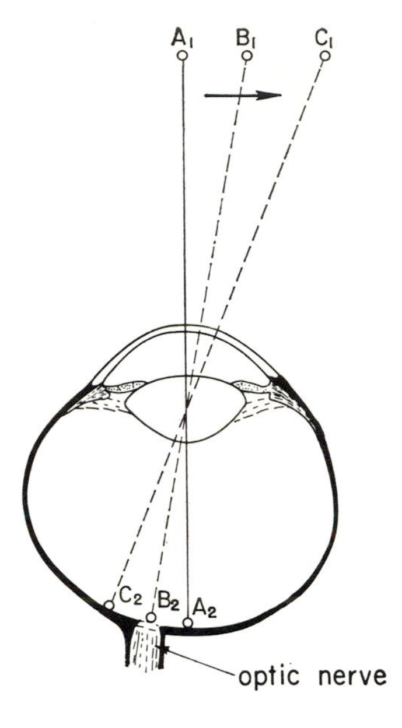

ring of light around the bright spot. The dispersion becomes greater if there is astigmatism or other optical distortion, but let us suppose that such defects have been corrected with proper spectacles. If now the observer uses the right eye, and the left eye is covered and the spot of light in Fig. 16 is moved from position A_1 to C_1 so that the image on the retina goes from A_2 to C_2, a sharp spot continues to be seen. But if a position B_1 is selected that will cause the image to fall on the blind spot B_2 (which is the place of exit of the optic nerve and lacks visual receptor cells), a new experience appears. Now the bright spot is no longer seen, and the whole eyeball seems lighted up with a diffuse, contourless glow. This glow represents the diffuse light within the eye that is always present when a stimulus is applied but is ordinarily inhibited.

To carry out this experiment properly the following steps are recommended. The left eye is covered and a very dim fixation point is located about 20 cm distant from the right eye. The fixation point is looked at continually, and it must be dim to prevent an inhibitory effect by itself. Then a second point of light, also 20 cm distant, is moved away from the fixation point slowly to the right, while the gaze is steadily maintained on the fixation point. When the second or stimulus light is about 45 to 50 mm to the right of the fixation point it disappears and the general glow is perceived as described above. The changes as the stimulus is moved are represented by the different sketches of Fig. 17.

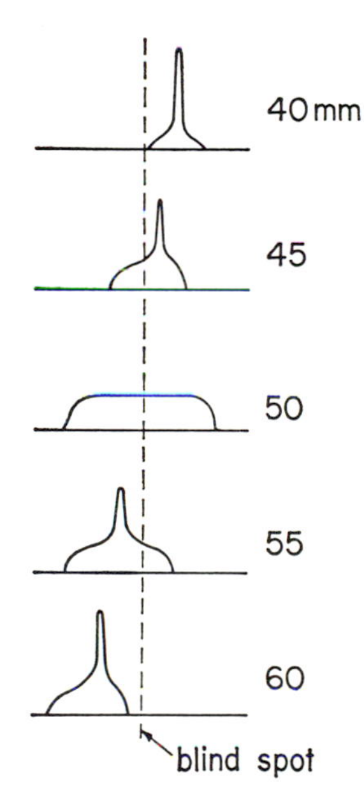

Fig. 17. The distribution of brightness in the visual field as a function of retinal region. A spot of light was moved across the region of the blind spot (from *A*2 to *C*2 in Fig. 16), giving the successive stages shown. The numbers on the right represent distances from the fixation point for a spot of light 200 mm from the eye. From *Acta Oto-laryngologica* [46].

This experiment gives about the same results for different colors and is largely dependent on the intensity of the stimulus spot. The effect has been found for all persons tested except one who had a disturbance of peripheral vision. In this observer, inhibition was present only near the center of the visual field, and even the slightest movement of the stimulus spot away from the fixation point gave a diffuse spot that expanded as the displacement of the stimulus was increased, reaching its maximum when the stimulus fell on the blind spot, as indicated in Fig. 18. The diffusion was symmetrical on right and left sides.

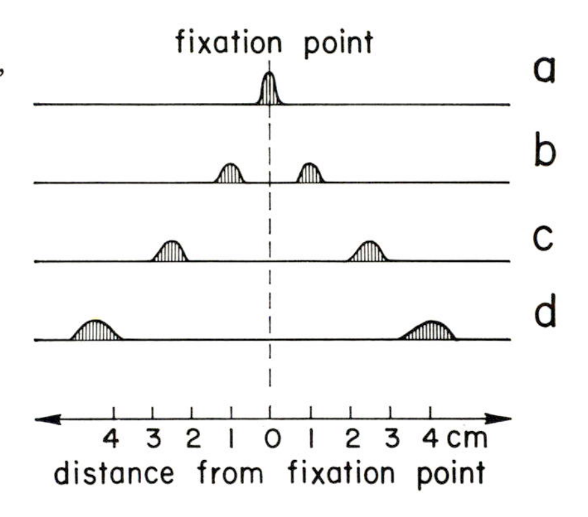

Fig. 18. Diffusion in a defective eye. Sketches *b*, *c*, *d* represent increasingly large rings of diffusion.

The question arises whether the reduction of the diffuse light around an image in ordinary vision is a peripheral or a central process. Anatomically considered, it might be a product of both. A test can be carried out as follows.

Let us arrange matters in the foregoing experimental situation so that the stimulus spot falls on the blind spot of the right eye, and then the left eye is opened so that the observer views the stimulus with both eyes at the same time. Under the conditions the stimulus spot will not fall on the blind spot of the left eye, so that now we have the situation represented in Fig. 19. Here *A* represents the normal image for the left eye, and *B* represents the diffuse effect caused by the image on the blind spot of the right eye. What is seen is pictured in *C*: there is a bright spot surrounded by a small amount of "fuzz." It appears that central inhibition has

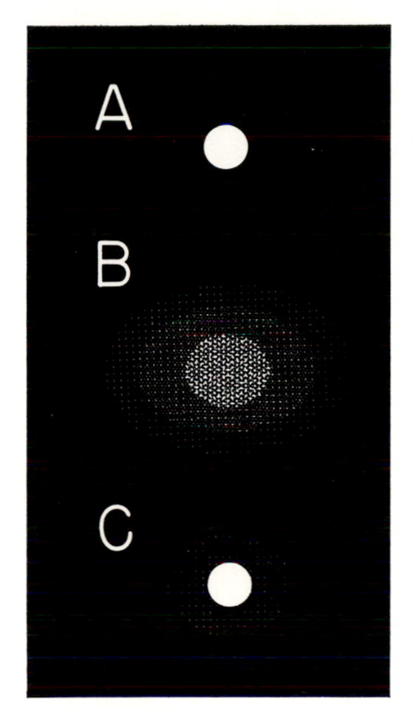

Fig. 19. A spot of light as seen under three conditions: *A*, with one eye when the image is on the retina, *B*, with one eye when the image is on the blind spot, and *C*, with both eyes when the image for one eye is on the retina and for the other eye is on the blind spot.

reduced the diffusion that the right eye alone sees as in *B*, but does not eliminate it altogether.

An estimate was made of the amount of decrease in the intensity of the fuzz brought about by central inhibition. The intensity of the fuzz when both eyes were used (as in *C*) was compared with that seen by the right eye only (as in *B*), and a grey wedge was used to reduce the intensity of *B* until it seemed equal to the fuzz in *C*. It was found that the reduction in intensity was about 200 times. It is clear that an inhibition that may arise in the central nervous system is by no means insignificant in amount.

II · THE INHIBITION OF SIMULTANEOUS STIMULI

When we look at a histological section of a sense organ with a large surface, we find nerve fibers going directly to the sensory cell, and in addition we find a large number of lateral fibers running parallel to the surface and connecting two or more end organs with one another. It was Held (1926) who showed that in the organ of Corti there are nerve fibers running from every hair cell more or less directly to the modiolus of the cochlea, and also other fibers running perpendicular to these along the basilar membrane, often connecting hair cells over nearly a half turn of the cochlea. Unfortunately these lateral connections often seem difficult to stain. The Golgi stain works well in young animals, especially prenatal ones, but perhaps the lateral fibers develop only at a later stage. The lateral fibers of the cochlea are represented in Fig. 20. At present it is still uncertain to what extent these fibers are involved in the immediate process of hearing. Recently Hartline (1949) and Ratliff, Hartline, and Miller (1963) studied lateral inhibition in the *Limulus* eye. This eye has a great advantage over many other sense organs because many of its fibers are embedded in fatty tissue and can easily be approached with electrodes. Figure 21 shows clearly how the ommatidia (the single units of this retina) are connected with the ascending nerve fibers. Also many of the lateral connections are visible.

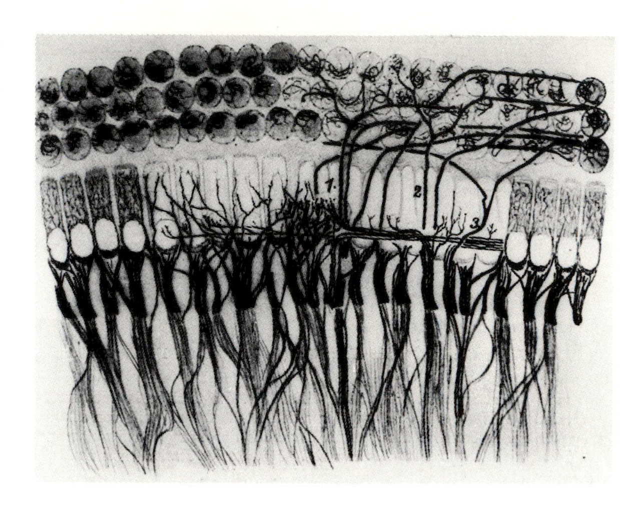

Fig. 20. Lateral interconnections between the hair cells of the cochlea. From Held (Bethe, A., *Handbuch der normalen und pathologischen Physiologie*, 11, *Receptionsorgane* I, 1926, p. 503; Springer-Verlag.)

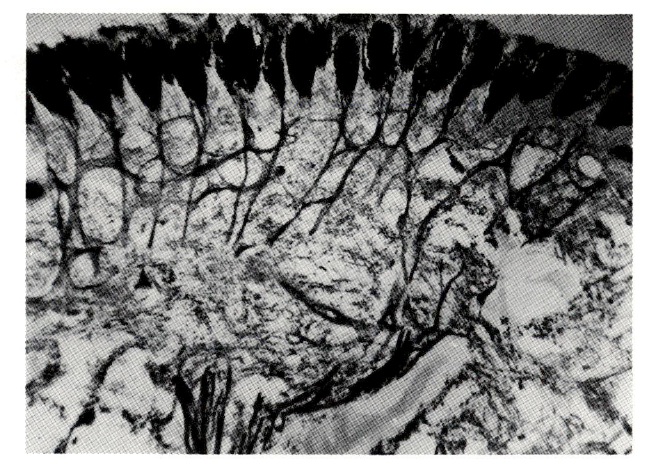

Fig. 21. Cross section of the *Limulus* eye. The black structures are ommatidia, whose nerve cords run downward to form the optic nerve. Other nerve cords produce lateral interconnections between the descending ones. From Hartline, Wagner, and Ratliff (*Journal of General Physiology*, 1956, 39, 651-673).

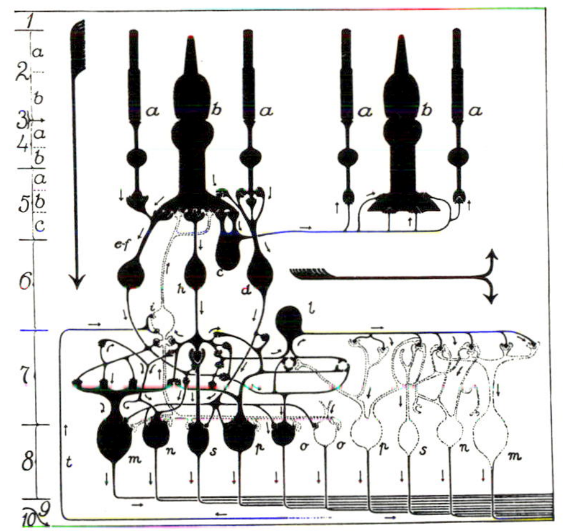

Fig. 22. Schematic cross section of the human retina showing the rods and cones and their nerve connections. Note the many lateral interconnections and feedback loops. After Polyak (*The Retina*, 1941, University of Chicago Press).

Figure 22, taken from Polyak (1941), shows in schematic form the lateral connections of the human retina. The arrows indicate the many directions that impulses may take and thus represent the diversity of connections. The pattern is as complex as many of the electric circuits used in computers. After the development of computer techniques it was widely believed that through these we would come to understand the circuitry of the retina. For my own part, however, I am unable to understand how this retinal circuitry fails to produce constant oscillation. It is difficult to imagine what type of reduction of feedback is used to avoid this oscillation. Even a simple feedback system requires many precautions to prevent its going into oscillation and to cause it to return to its original equilibrium condition after a stimulus has ceased to act upon it.

Fig. 23. Drawing to represent different levels of lateral interaction in the skin, and the resulting sharpening effect. From Ruch and Fulton (*Medical Physiology and Biophysics,* 1960, Fig. 183; W. B. Saunders Co.).

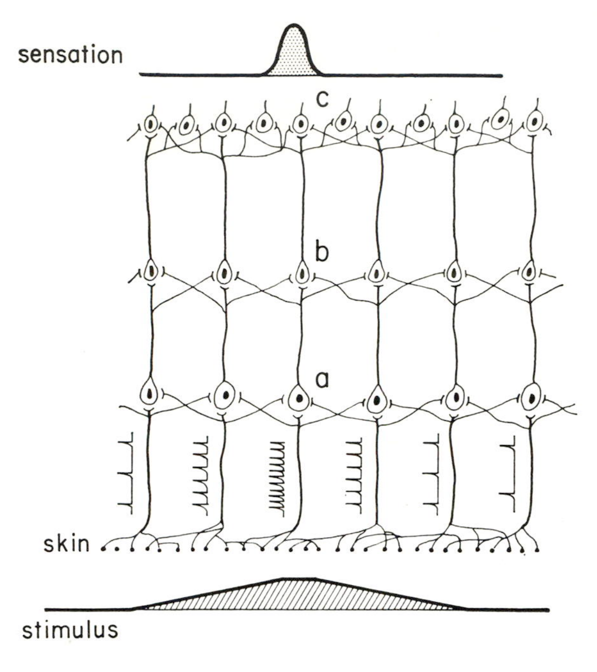

For present purposes, a simple scheme is sufficient to illustrate the problem. The small dots near the bottom of Fig. 23 represent a system of receptors on the surface of the skin. The variations of a stimulus are indicated by the lowermost graph with a maximum in the middle. There are lateral connections between the receptors and nerve fibers that run upward to the first-order ganglion cells in row *a*. Alongside each of these nerve fibers is a representation of its pulse rate: the number of spikes for a certain unit of time is shown. Adrian in 1928 found that the discharge rate of a sensory nerve fiber increases with the magnitude of the stimulus acting on its end organ. In the two fibers on the far left and right sides the stimulation is minimal and the discharges represent spontaneous activity. Toward the middle the stimulus intensity increases and so also does the discharge rate.

From the first layer of ganglion cells we go to a similar level above, shown at *b*, and from there to a still higher level, *c*. As far as we can now discover, the result of these lateral interconnections is to reduce or "funnel" the laterally spreading stimulation to a progressively localized section of the neural pathway, as indicated by the sketch of the sensation at the top of the figure.

This simple scheme immediately brings up the question of the kind of frequency sensation that will be produced by a stimulus whose amplitude tapers away from a central maximum as shown in this figure. It is apparent that the lateral inhibition that occurs in sense organs is not the straightforward kind that we find in muscle systems. The lateral inhibition of sense organs is actually a funneling action that inhibits the smaller stimulus effects and collects the stronger effects into a common pathway.

I consider the human skin to be a favorable place to investigate this funneling action. The endings are at the surface and no surgery is required to disclose them (as is the case for the cochlea). Also the skin contains various kinds of sense organs, such as pressure receptors, vibratory receptors, warm receptors, and cold receptors. We can go from the skin to the inside surface of the mouth and to the tongue, where we can observe not only the ectodermal but also the endodermal nerve supply. Under some conditions the skin exhibits regeneration, and then an even greater variety of sensory conditions can be investigated.

The skin presents one problem: it is difficult to anesthetize with topical application of substances. The use of endophoresis to force the anesthetic into the skin is perhaps the best method. A fortunate circumstance, however, is that the end organs of the skin can be altered considerably in sensitivity by warming or cooling. Then their activities may be compared with the normal state.

The network of Fig. 23 contains only ascending and lateral connecting fibers. There are probably structures, like the *Limulus* eye, that are as simple as this, but in general there is also a descending network as well.

In a crude way we can distinguish four types of inhibitory pathways, as illustrated in Fig. 24. At *A* is shown the simple form of lateral inhibition that has just been described. At *B* is shown a forward type of inhibition, at *C* a backward type, and at *D* a form of central inhibition. In human sense organs there is an interplay of all these types of interaction.

The question arises whether inhibition is the correct term for this whole series of phenomena. With our limited knowledge of inhibitory phenomena in single neural pathways it is difficult to decide on the proper category. In addition to inhibition we might consider the term "sensory distortion," or even "disinhibition"—because it may happen that normally the neural pathways are blocked, and the effect of a stimulus is to remove the blockage. I prefer to call the effect "funneling," because even for central inhibition there is usually an increase in the magnitude of the sensation referred to certain places.

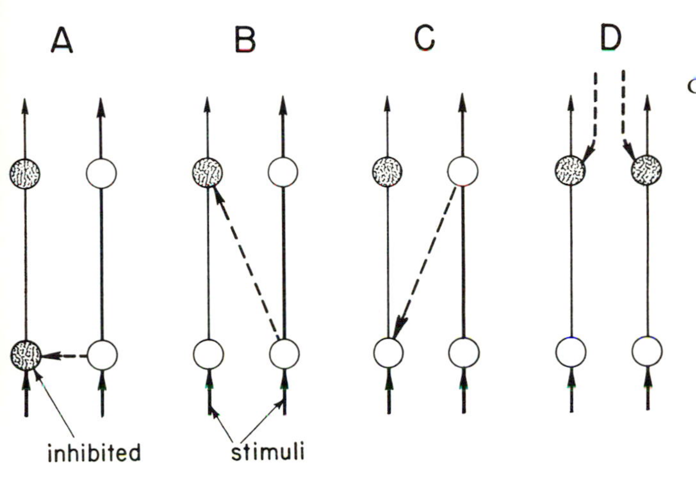

Fig. 24. Different types of inhibition. After Reichardt and MacGinitie (*Kybernetik*, 1962, 1, 155-165).

Simultaneous inhibition on the skin

As already pointed out, the easiest place to observe the different forms of inhibition is on the skin. Here is where I first observed it, because the observation requires very simple equipment. Two test hairs (known as von Frey hairs from his early use of them; see von Frey 1926, 1929) are sufficient to show the effects of lateral inhibition over the skin surface. These hairs make it possible to produce pressures of constant value, as pictured in Fig. 25. By touching the surface of the skin with a hair, while hold-

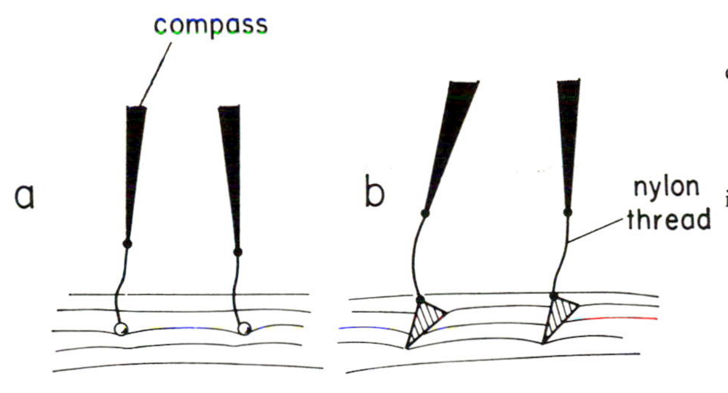

Fig. 25. Von Frey test hairs. At *a* are the conventional type of tips. Improved precision is obtained by the use of nylon threads and triangular metal tips, as in *b*. From *Journal of the Acoustical Society of America* [87].

ing the handle in which it is fixed with the free hand, a direct type of pressure is produced. To observe lateral inhibition, two test hairs are attached to the legs of a pair of compasses, and the legs are separated to alter the separation of the points.

Two test hairs were applied to the palm of the hand of an observer as indicated in Fig. 26, and he was asked to describe the distribution of the direct pressure sensation at the point of application and around it. This observation is difficult because we are accustomed to note only the maximum point of a sensation and to report this as the magnitude. However, observers can easily be trained to observe the magnitude variations along a surface. It seemed to me that the easiest way to proceed was first to give the observer some preliminary training by presenting him with a well-defined pattern of light intensities and asking him to make a drawing of the effects upon the retina. After being given various visual patterns, he was able to observe the local differences in pressure sensations and to express them in an arbitrary scale of magnitudes. No great precision can be obtained by this method, but some of the phenomena under discussion can be observed.

Figure 26 shows the pattern of sensation when only one compass point is pressed against the skin, and also the effects of two points that are applied with equal pressure but varied in their distance from one another from 1.5 to 3.5 cm. At the 1.5-cm distance the perceived pattern is still unitary but is spread out laterally, and also is a little greater in magnitude at its maximum.

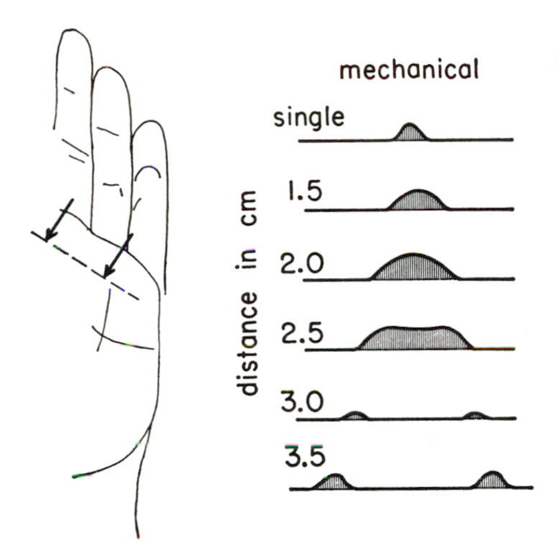

Fig. 26. The local distribution of sensory magnitude for mechanical stimulation with two points. From *Journal of the Acoustical Society of America* [87].

The magnitude and the width of the pattern grow greater as the separation of the points is increased, until at a certain distance two points are perceived. At the same time, as shown for a separation of 3.0 cm, the magnitude of the sensation decreases sharply. Sometimes, with careful adjustment of the two stimulus magnitudes, it is possible to cancel the sensation completely, at least for a brief time. At a little larger distance the two stimuli fall apart, there is no more lateral interaction, and the two points feel like two independent stimulations. The disappearance of sensation at a certain distance between the two points seemed to me to be the most striking demonstration of lateral inhibition.

The complete phenomenon is not found for very weak stimuli. For these, as the right-hand portion of Fig. 27 shows, there is only summation. Strong stimuli are required to exhibit lateral inhibition. These tests were performed mainly on the upper arm.

Fig. 27. The distribution of sensory magnitude as a function of stimulating pressure. From *Journal of the Acoustical Society of America* [87].

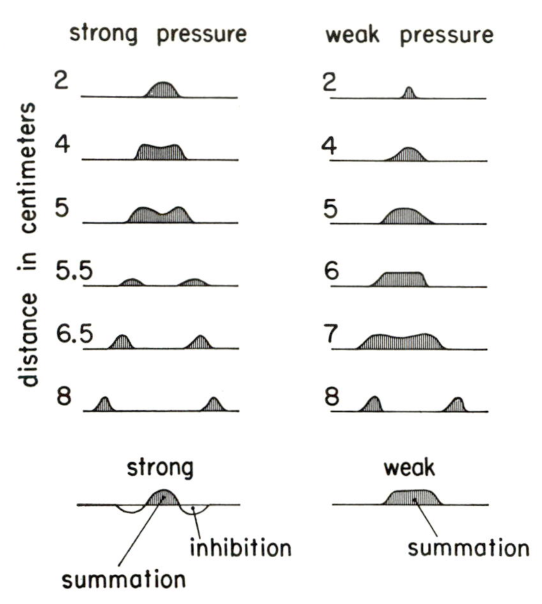

This kind of inhibition depends also upon the temperature of the skin surface, as Fig. 28 portrays. The observations of temperature effects were made primarily to discover whether lateral inhibition is a peripheral or a central phenomenon. The results indicate that there is probably a combination of peripheral and central effects. Both summation and inhibition are involved, but conditions such as temperature affect the relative roles of these two processes.

It seems to me that the patterns of sensation produced by two-point stimulation of the skin may be accounted for by assuming first that every stimulus produces an "area of sensation," as shown in Fig. 29. This area is surrounded by an area of inhibition and of decreased sensitivity, so that every sharply localized stimulus

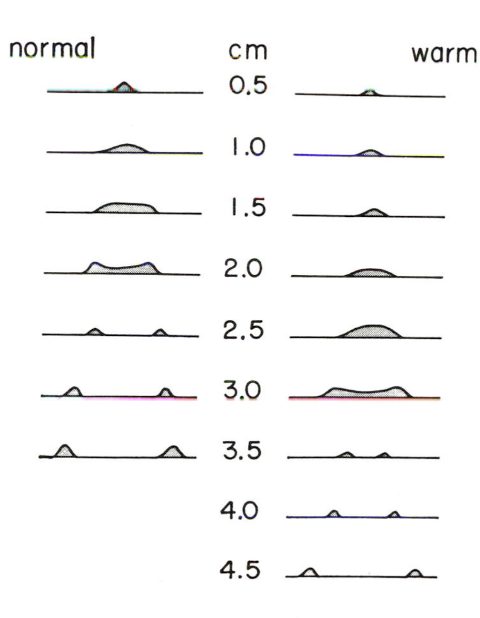

Fig. 28. The distribution of sensory magnitude as a function of skin temperature. From *Journal of the Acoustical Society of America* [83].

produces two effects. At present the area of inhibition is of primary concern because it allows us to describe many strange variations in the distribution of sensation magnitude. The inhibitory area may change strikingly with the temporal or spatial pattern of stimulation. Some of these effects are shown in Fig. 30.

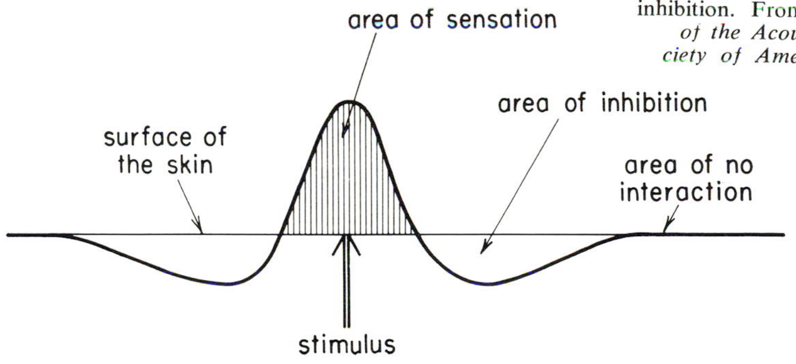

Fig. 29. The area of sensation produced by a local stimulus, and its surrounding area of inhibition. From *Journal of the Acoustical Society of America* [87].

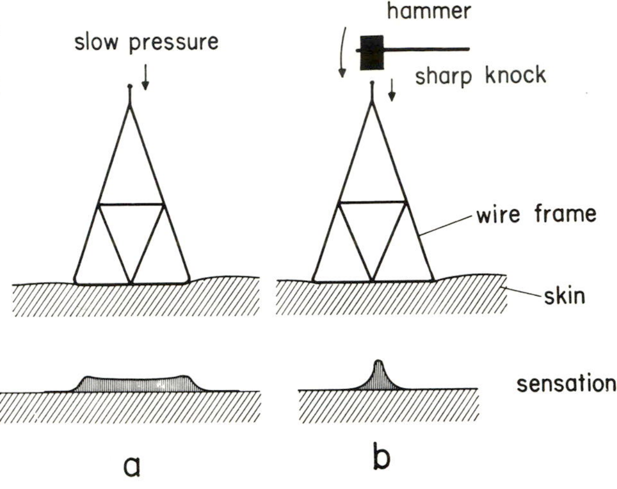

Fig. 30. The distribution of sensation as a function of the temporal pattern of stimulation. From *Journal of the Acoustical Society of America* [74].

If the lower edge of a wire frame is slowly pressed against the skin of the upper arm, observers will describe the pattern as rather uniform in magnitude, as seen at *a* in Fig. 30. If, however, the frame is struck sharply with a hammer, as shown at *b*, the lateral spread of the magnitude is slight and the sensation is more localized. The sharpness of localization increases as the striking of the frame is made more abrupt. This observation seems to indicate that inhibition is strong for stimuli with rapid onset, producing a great amount of "funneling." In agreement with this conclusion, a series of sharp blows produced by driving a large frame with an electrodynamic vibrator gives a smaller lateral spread when the vibrator is actuated by clicks (upper part of Fig. 31) than when it is actuated by sinusoidal waves of the same periodicity (lower part of this figure). Incidentally, it is good training to have observers make drawings of the distribution of perceived magnitudes for these types of stimulation.

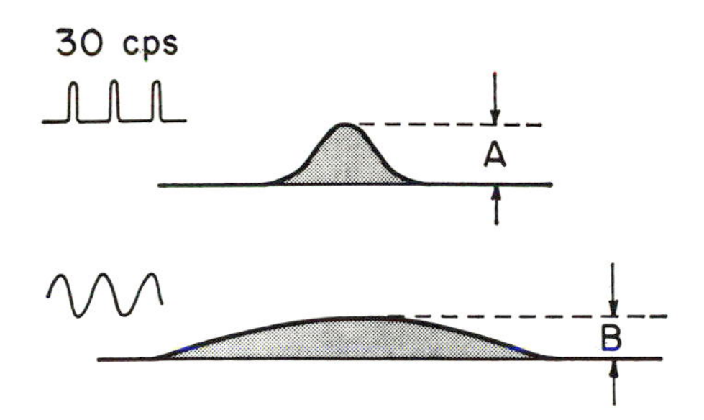

Fig. 31. A comparison of the effects of click and sinusoidal stimulation. From *Journal of the Acoustical Society of America* [83].

The decrease in the lateral spread with an increase in the frequency of stimulation is a universal phenomenon. As shown in Fig. 32, it holds for mechanical vibrations, electrical stimulation, and hearing.

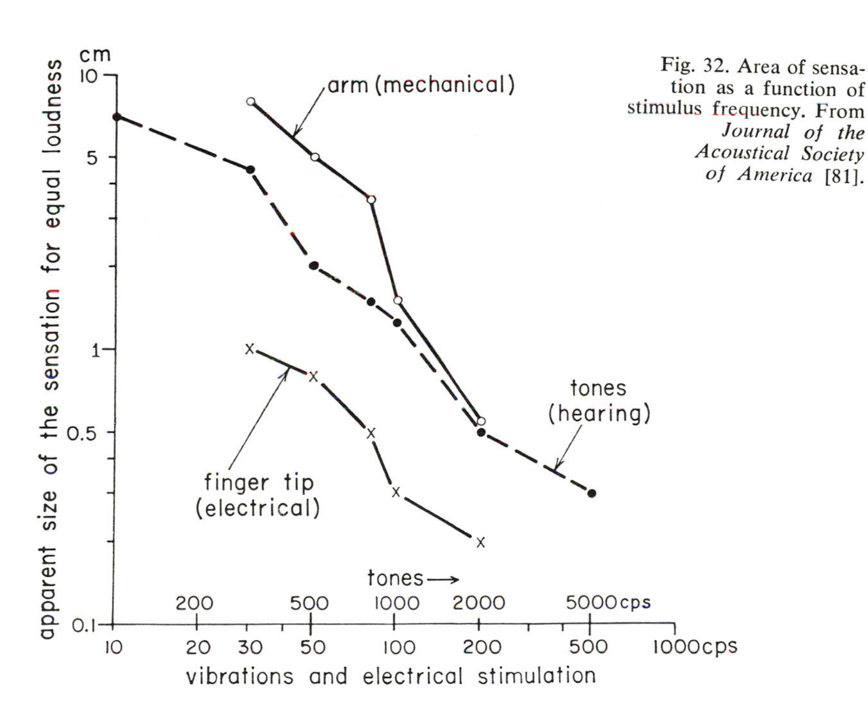

Fig. 32. Area of sensation as a function of stimulus frequency. From *Journal of the Acoustical Society of America* [81].

Apart from the change in the lateral spread for two-point stimulation, there are some very complicated phenomena when the vibratory "pitch" for two vibrating points is compared with the "pitch" for a single point. Figure 33 gives the results of equating the "pitch" of two points at varying distances with that of a single point. The separation of the two points clearly has an effect. At present I have no explanation of this phenomenon. It is obvious that these effects of inhibition depart widely from our beloved input-output concepts.

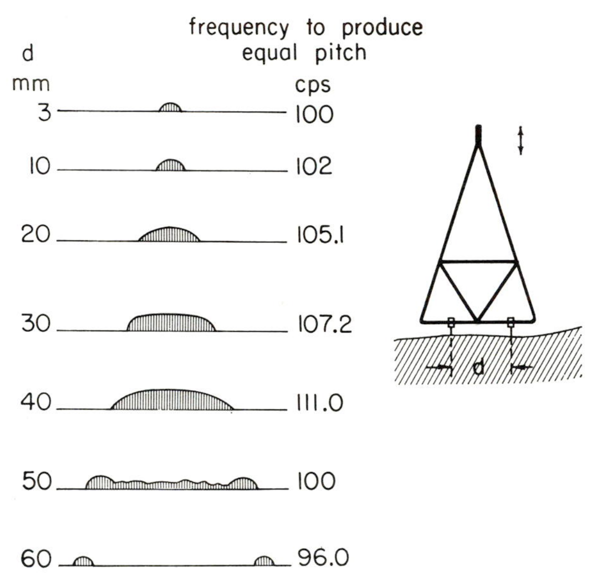

Fig. 33. Vibratory "pitch" as a function of the separation of two stimulating points. From *Journal of the Acoustical Society of America* [85].

Frequency inhibition

In addition to changes in the magnitude patterns as just described, local stimulations on the skin surface produce other effects that I wish to call "frequency inhibition." These effects

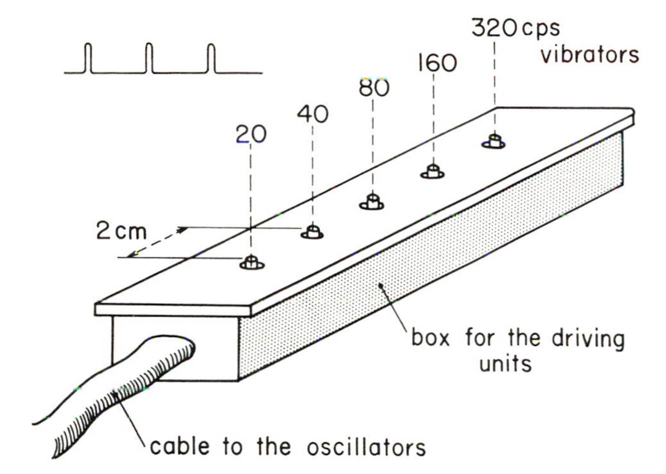

Fig. 34. A series of vibrators operating at different frequencies, used in the study of inhibitory interaction. From *Journal of the Acoustical Society of America* [81].

arise from experiments like the following. A set of small vibrators was placed in a long box, as shown in Fig. 34. Each vibrator could be adjusted in amplitude, and they were driven by means of sharp clicks at various frequencies as shown. Though the frequency increased from left to right along the array, the amplitude was adjusted so that for each vibrator felt by itself it seemed the same. The array was placed with the vibrators acting on the upper arm, and held carefully. When all the vibrators were switched on at once, there was the impression of a single vibratory "pitch," which was that of the 80-cps vibrator presented alone. Also the sensation produced by the array was felt as coming only from the middle vibrator, as illustrated in Fig. 35. From this experiment it is clear that there is a well-defined inhibition of "pitch." When the adjustments are made precisely, the frequencies of 20, 40, 160, and 320 cps are simply not perceived. This form of inhibition I regard as important in explaining certain electrophysiological phenomena, such as those of Fig. 23, in which the discharge

Fig. 35. A representation of the sensory effects of stimulating the skin in the manner indicated in the preceding figure. From *Journal of the Acoustical Society of America* [81].

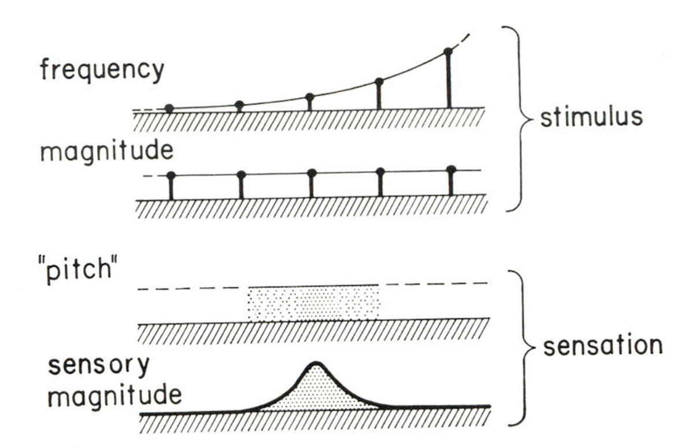

frequency of certain nerve fibers (those at the periphery) is smaller than that of those in the interior of the series. The effect is like that just described, except that in the neural network the discharge of lower frequency is always inhibited.

The time of onset of frequency inhibition was studied after a particularly careful equating of the perceived magnitudes of the test array shown in Fig. 34. It was found that this inhibition develops slowly, as *a* in Fig. 36 indicates. In this instance the inhibition was present only after 10 to 20 seconds. The rate of onset was greatly increased if the magnitude of the 80-cps vibrator was slightly increased. As part *b* of this figure shows, an increase of 30 per cent in the magnitude of this middle vibrator reduced the onset time to a few seconds.

The interaction of two-point stimulation exhibits the summational and inhibitory effects also in the relation between perceived magnitude and the amplitude of the vibration. The frame with two points, as shown in Fig. 33, was vibrated at a rate of 150 cps, and applied to the upper arm. When the distance between the points was 1 cm and the perceived magnitude

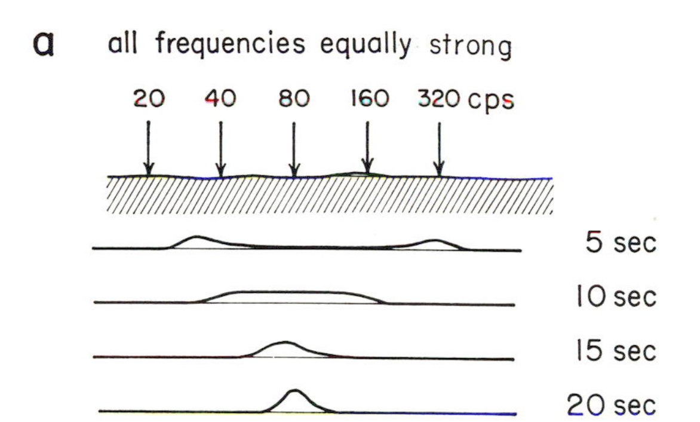

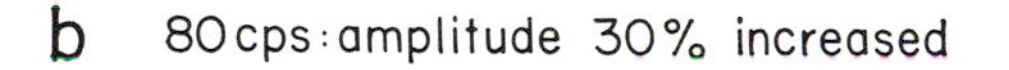

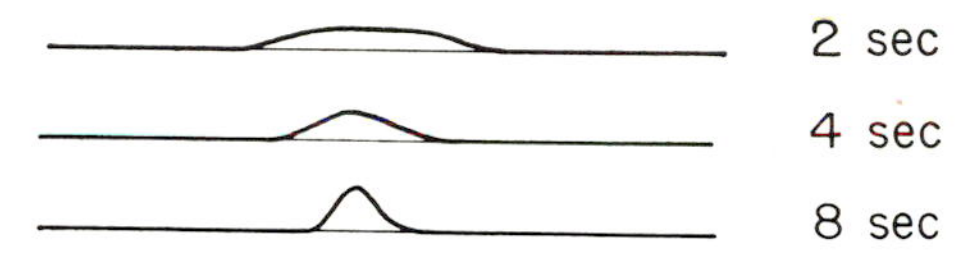

Fig. 36. The development of lateral inhibition through time. From *Journal of the Acoustical Society of America* [81].

was compared with that produced by the frame itself (with the two points removed) it was found that near threshold the perceived magnitude of the two-point stimulation rose relatively rapidly as a function of vibratory amplitude—more rapidly than that for the line of stimulation produced by the frame, as may be seen in Fig. 37. When the distance between the two points was increased to 4 cm, the rate of increase with stimulus intensity was about the same as for line stimulation. However, when the distance between the two points was further increased to 8 cm the rate of increase again was greater for the two points than for line stimulation. Clearly the relation between perceived magnitude and stimulus intensity changes with the spatial configuration.

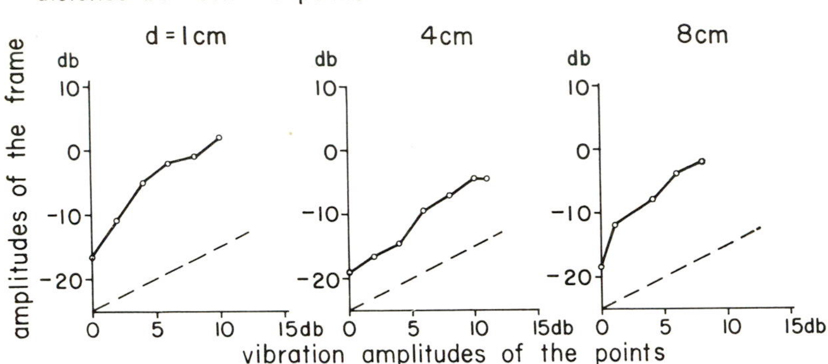

Fig. 37. A comparison of the perceived magnitudes of vibratory sensations when produced by two vibrating points (at three different separations) with the perceived magnitudes produced by a vibrating frame. The broken lines show the slopes of the curves that would have been obtained if the perceived magnitudes increased at the same rate for both conditions. From *Journal of the Acoustical Society of America* [83].

Similar results were obtained when the difference limen for perceived magnitude was determined for two-point stimulation with different separations of the points, as represented in Fig. 38. These results were obtained by asking an observer to press a button and hold it until he felt a vibration and then to release it, then to press it again when the vibration was no longer perceived. As the figure shows, the changes required to go back and forth between perceived and unperceived were small when the separation between the points was 0.5 cm, became larger when the separation was 4 cm, and became smaller again when the separation was 10 cm. This is a rough method, and I regard it as a means of determining the difference limen near the threshold. The feature of interest is that the interaction of two points is sufficient to improve the over-all sensitivity of a group of receptors.

A striking variation of this experiment consisted in applying adhesive tape to the arm, as shown in Fig. 39, so that the vibrating frame acted simultaneously upon an increased number of end organs. In this situation the size of

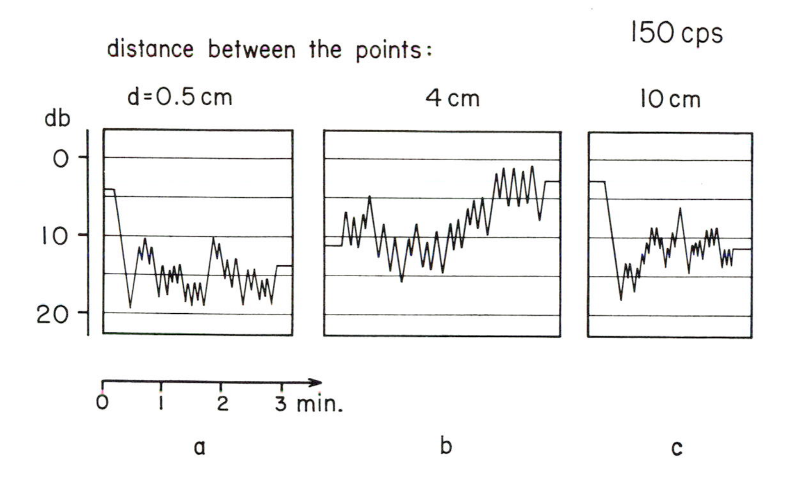

Fig. 38. Size of the difference limen near threshold for different separations of the stimulating points. From *Journal of the Acoustical Society of America* [83].

the difference limen increased markedly, as *a* in Fig. 40 shows. Thresholds and difference limens seem to be closely connected. The smaller the number of nerve endings involved in perception the smaller is the difference limen for perceived magnitude of vibration. By the use of a needle a very small difference limen was obtained, as shown in *b* of Fig. 40.

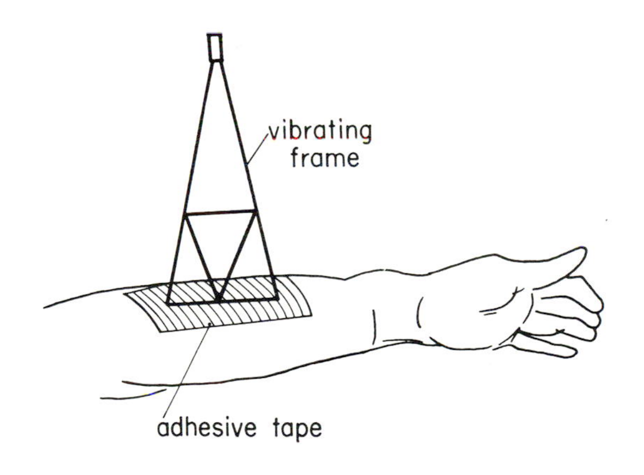

Fig. 39. Method of studying areal summation. From *Journal of the Acoustical Society of America* [83].

Fig. 40. Size of the difference limen for a large and a minimal area. From *Journal of the Acoustical Society of America* [83].

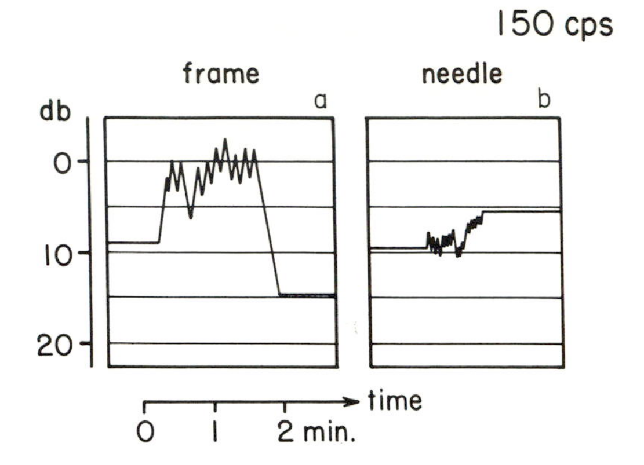

Change in localization resulting from the inequality of two stimuli

In hearing it is well known that a hummed tune increases in loudness and moves from the middle of the head to one side when we lightly close one ear with a finger. The increase of sound pressure in the closed ear can be no more than 4 db, yet this is sufficient to produce a complete lateralization of the tune to this side. Thus a slight loudness imbalance can cause large changes in the localization of a stimulus. This phenomenon is the basis of the Weber test used by otologists to detect the difference between the two ears in unilateral middle-ear defects.

For the skin, as has been shown in the two-point stimulation situation, two stimuli when only a short distance apart add their effects to produce a common sensation. It is easy to displace the center of this sensation by having the two magnitudes unequal. In general the combined sensation is displaced toward the stronger stimulus and the weaker is inhibited.

As shown in Fig. 41, two vibrators acting on

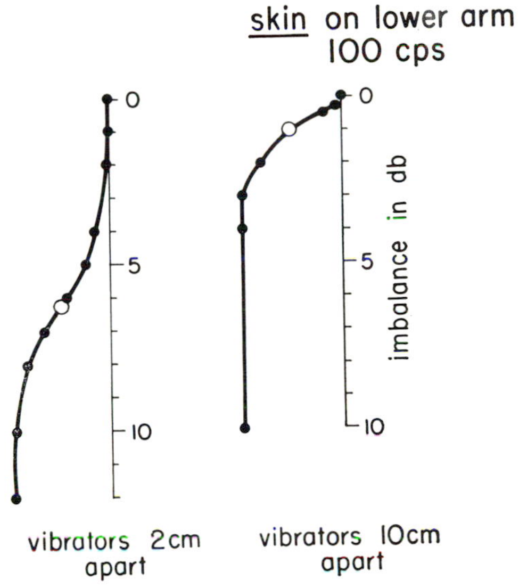

Fig. 41. Effect of imbalance between two vibrating points for two different separations of the points. From *Journal of the Acoustical Society of America* [87].

the skin of the upper arm, when 2 cm apart, cause a displacement of the common sensation that is all the way to one side when the vibrator on that side is 10 db stronger than the other. If the separation is 10 cm, then 3 db is sufficient to produce the lateralization. In this range of distances the movement from the center as one stimulus is changed relative to the other is continuous. But if the stimulus separation becomes as great as 20 cm the displacement is no longer smooth and there is a jump from the middle position to one side. It seems that if the two stimuli are close together their neural patterns are joined by numerous lateral connections, so that an interaction produces a continuous transition from one place to another. But when the two stimuli are far apart the connections between the patterns are few, and the interaction is an all-or-none affair producing a jump from one condition to another.

Fig. 42. Effect of imbalance between two vibrating points as a function of skin temperature. From *Journal of the Acoustical Society of America* [87].

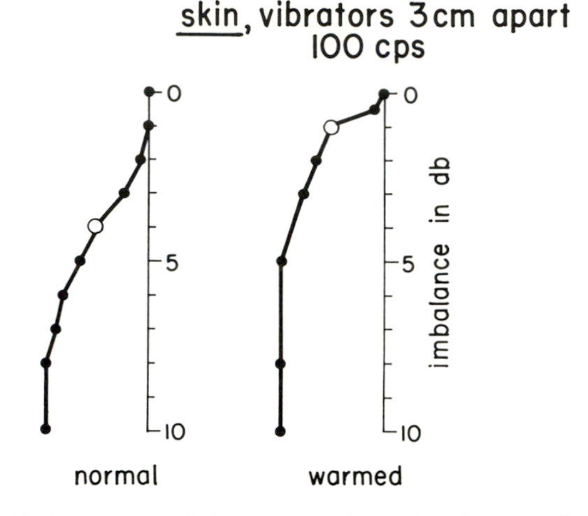

It is supposed that warming the skin surface decreases neural activity in both the ascending and lateral nerve connections. It follows that warming produces effects like those given by increasing the separation of two stimuli. The evidence of Fig. 42 supports this hypothesis.

It is an advantage of research on the skin that the end organs can readily be caused to change their sensitivity by warming, cooling, rubbing, and anesthesia. By these procedures it is possible to differentiate in some degree the operation of peripheral and central structures. The experiments already mentioned indicate that neural connections are modified by warming, cooling, etc. The phenomena then exhibit all-or-none characteristics, as is commonly observed when we substitute a simple neural unit for a complex system.

A change in localization as a function of the magnitude relation between two stimuli is found not only for the two-point situation but also for a continuous distribution of stimulation. In

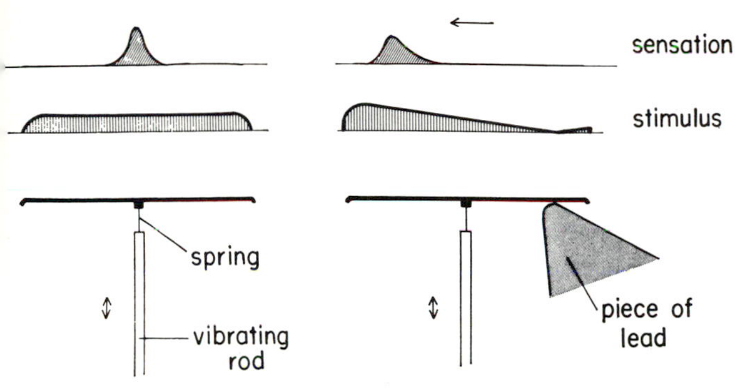

Fig. 43. The effect of stimulus pattern on the location of a perceived maximum. A mass in contact with the vibrating tube modifies the pattern of stimulation. From *Journal of the Acoustical Society of America* [74].

Fig. 43 we see a small tube fixed to a vibrating rod that can be applied to the skin of the upper arm so that the amplitude of vibration is equal along its length. In this case the sensation is localized at the middle, and there is little lateral spread. If now one end of the vibrating tube is touched with a piece of lead (as shown at the right of the figure) the maximum of vibration moves to the other end and so also does the sensation.

An even more interesting case is one in which the amplitude of vibration is kept constant but the stimulation is done in a region like the hand where the sensitivity changes. Figure 44 shows how the sensitivity of the skin increases as we approach the fingertips. There are changes not only in sensitivity, but in all other properties of vibratory sensations, as shown in Figs. 44 and 45.

If we stimulate the four fingers and the palm as indicated in Fig. 46, the location of the vibratory sensation moves from the fingertips to the palm as the stimulus intensity is increased (Fig. 47). Here is a particularly interesting phenomenon in which inhibition is involved, and one that deserves careful analysis.

Fig. 44. Character of sensation as a function of density of innervation. From *Journal of the Acoustical Society of America* [81].

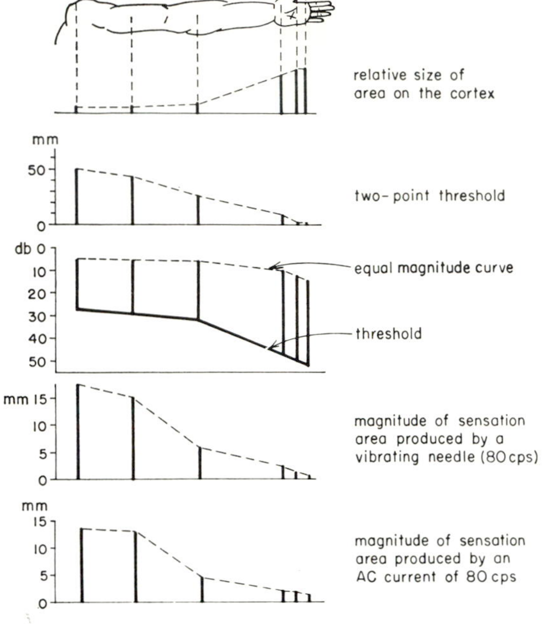

Fig. 45. Further variations of sensation as a function of innervation density. *From Journal of the Acoustical Society of America* [81].

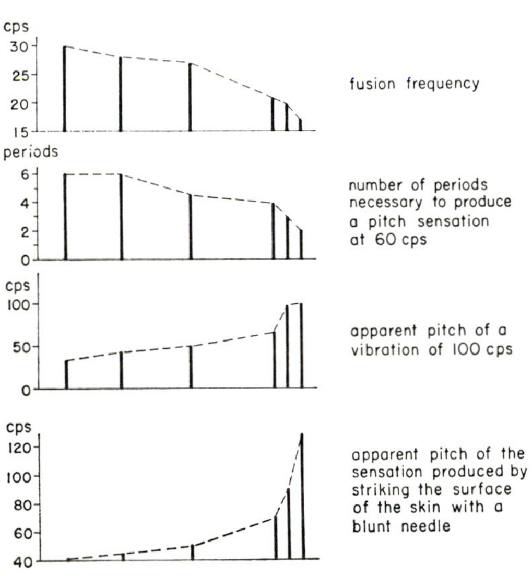

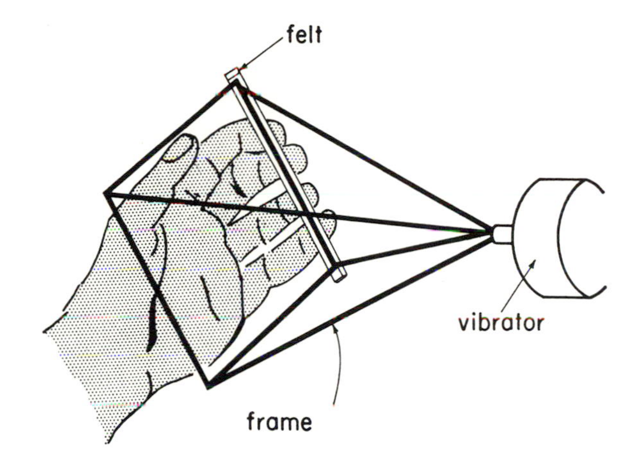

Fig. 46. Method of simultaneously stimulating two regions of the skin. From *Journal of the Acoustical Society of America* [87].

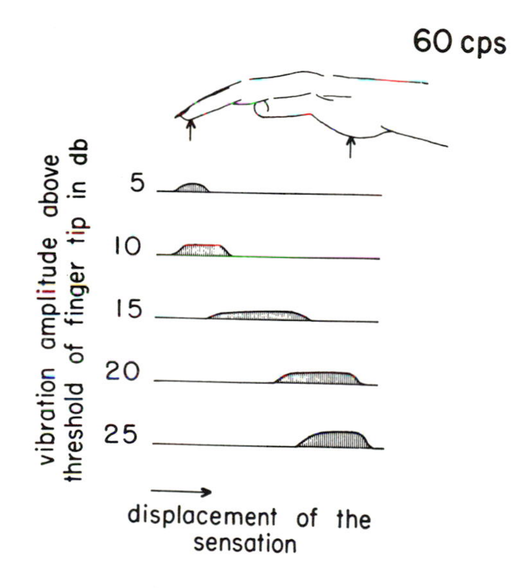

Fig. 47. Effect on localization of the amplitude of stimulation of two regions of different innervation density. From *Journal of the Acoustical Society of America* [87].

The reason for the shift in localization with change of stimulus intensity lies in the slower increase in sensed magnitude, as the stimulus intensity is raised, for the fingertips than for the palm. The fingertips with their greater sensitivity have a lower rate of increase with intensity, hence the regions toward the palm gain relative to the fingertips as the stimulus intensity is raised.

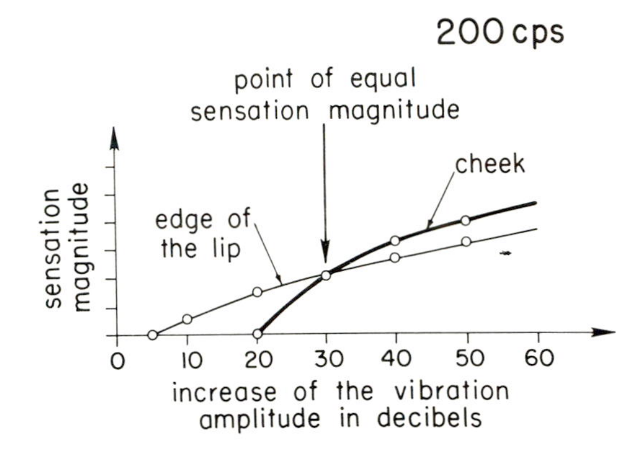

Fig. 48. Form of increase in sensory magnitude as a function of vibratory amplitude, for two regions of different density of innervation (lip and cheek). From *Journal of the Acoustical Society of America* [87].

Figure 48 shows the same effect in the region between the edge of the lip and the cheek. The edge of the lip is the most sensitive, and areas toward the cheek are progressively less sensitive. As the amplitude of vibratory stimulation increases the sensation is localized first at the lip and then farther and farther toward the cheek. A certain magnitude of stimulation gives a point where the sensation is localized midway between lip and cheek, and here the two extreme sets of endings may be considered as giving equal sensory magnitudes.

A third method of studying the effect of differences in sensitivity on localization is to use two similar skin areas but to reduce the sensitivity in one of them by warming it. A method of doing this experiment is shown in Fig. 49, together with the results. When the surfaces of a normal and a warmed finger are stimulated with the same vibrator, the sensation is first localized on the cold finger at low intensities of stimulation and moves to the warm finger as the intensity is increased. These experiments show convincingly how complicated the processes of "funneling" are.

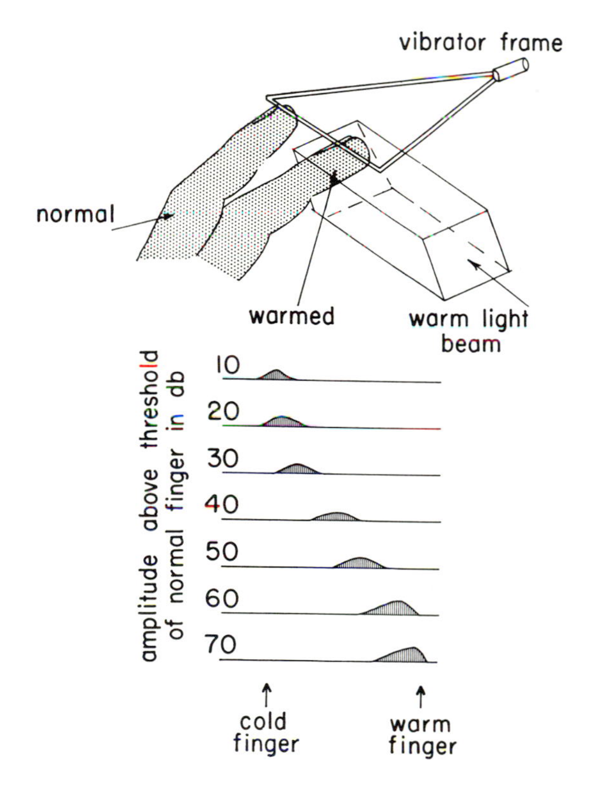

Fig. 49. Effects of altering the sensitivity of the skin by the application of heat. From *Journal of the Acoustical Society of America* [87].

It is now an open question how the brain discriminates the magnitudes of sensation. There are several possibilities. One is that the increase in magnitude of sensation is proportional to the number of neural discharges per second received in a higher center. Another possibility is that the number of stimulated nerve fibers increases with stimulus magnitude. Both of these assumptions are well supported by electrophysiological observations.

It is possible also, however, that there is a transformation of stimulus magnitude into a displacement of neural activity of the type just seen in the skin experiments. In addition to the form of displacement described here, there are other systems that might produce a displacement of sensation. One of these will be examined.

In Fig. 50 is shown a pair of bars that may be set in vibration in different ways. Bar *A* vibrates laterally as indicated by the arrows. Bar *B* rotates around the axis shown at its right end. A stroboscopic method was used to ensure that the two bars vibrated in the ways specified. Now the lower arm was placed between these two bars, with the hand pointing to the right of the figure. The bars were set in vibration, with the amplitude of *B* kept constant and that of *A* continually decreased. The locus of the sensation, which was initially at the left of the figure, i.e., on the arm, then moved progressively to the right, toward the hand.

This observation was made also when, instead of continuous vibration of the bars, they were actuated with two sharp clicks each second. This method avoids an adaptation of the skin. Figure 50 also shows the results obtained with three

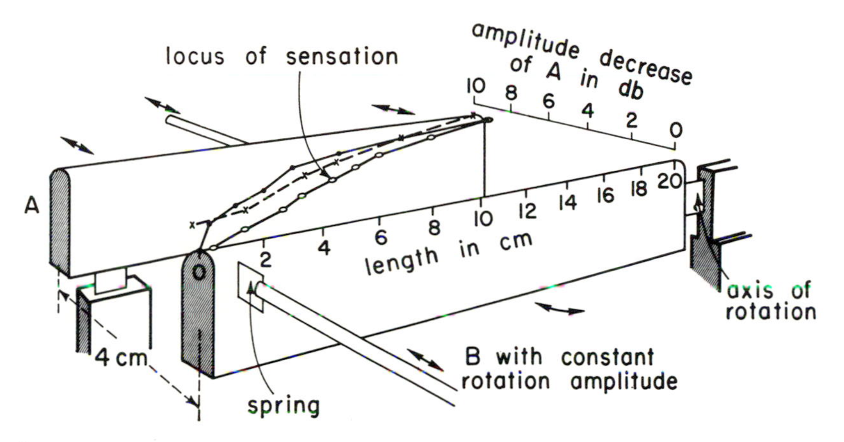

Fig. 50. Changes in the locus of sensation produced by variations in the relative vibrational amplitude of two vibrating bars. The three curves represent results for three different subjects.

observers. It is surprising that an amplitude decrease of bar *A* can displace the sensation from the elbow to the hand, a distance of nearly 20 cm. It is a question whether processes as determinative as these may be found in higher centers of the nervous system.

The Mach bands

Lateral inhibition has long been known in vision. Its role in simultaneous contrast and in color contrast has been intensively studied. The literature in this field is enormous and more than I can cover in this lecture series. I shall omit a discussion of simultaneous contrast and deal only with the Mach bands and related phenomena. These two kinds of inhibition are not necessarily the same, or at least are different quantitatively, because simultaneous contrast involves a much wider spread of lateral action than the inhibition concerned in the Mach bands.

Fig. 51. The disk used by Mach. From *Sitzungsberichte der Akademie der Wissenschaften in Wien, math.-nat. Cl.* (1865, 52, pt. 2, 303-322).

Fig. 52. The mode of appearance of the disk shown in the preceding figure when it is rotated rapidly. From *Sitzungsberichte der Akademie der Wissenschaften in Wien, math.-nat. Cl.* (1865, 52, pt. 2, 303-322).

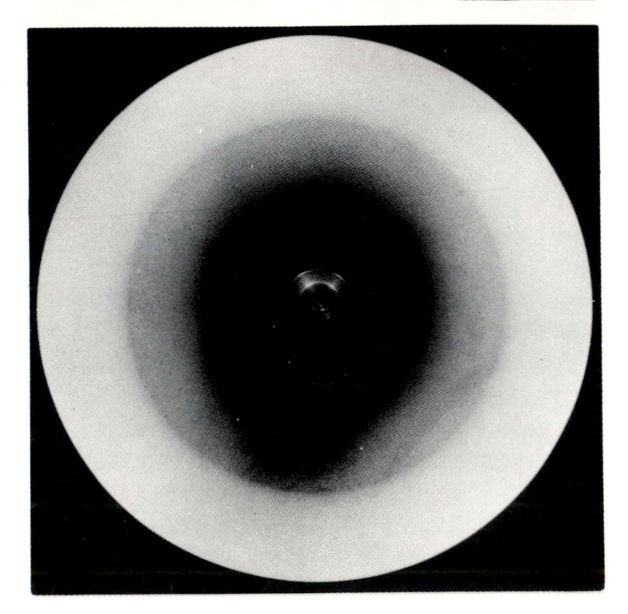

The Mach bands are independent of movements of the eyeballs, for they can be seen when the retinal image is optically stabilized (Riggs, Ratliff, and Keesey, 1961).

Mach, a physicist much interested in psychological phenomena, first observed the bands that are known by his name in 1866 and 1868. He obtained them first by rotating a disk like that of Fig. 51. When the speed of rotation is great enough to eliminate flicker, a pattern like that shown in Fig. 52 is obtained. The general expectation was that this uneven pattern of light intensity would produce summation in some regions and a sort of overshoot, for instance at the outer edge. But Mach found that in addition to the overshoot there was a diversity of sensory magnitudes and the formation of a black ring, which represents an undershoot. The black ring was often darker than the black of the paper from which the pattern had been made.

It was Mach's genius to realize that this undershoot was an important new phenomenon. It was a significant discovery of lateral inhibition in vision. If we represent the distribution of light intensity and of perceived brightness as in Fig. 53, we find that at every place where the light intensity suddenly begins to accelerate (i.e., to increase its rate of increase) there is the formation of a dark line. On the other hand,

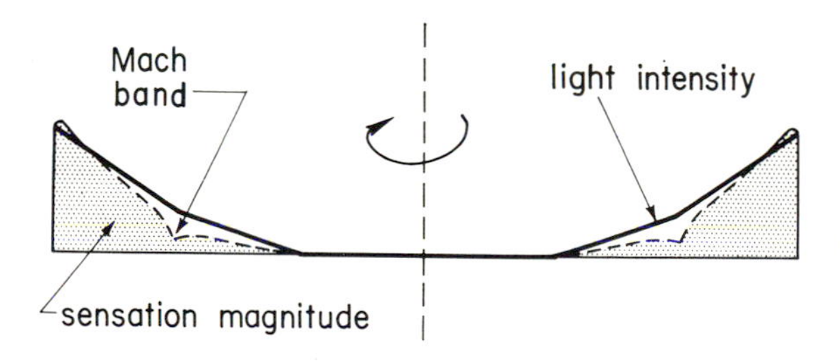

Fig. 53. Effects of a discontinuity in the pattern of stimulation on the pattern of sensation. From U.S. Army Medical Research Laboratory, Fort Knox, Kentucky, Report No. 424 [93].

in the places where the light intensity begins to decelerate there is an overshoot with the formation of a white line. These lines are the Mach bands. The dark bands are blacker than the original stimulus surface, and represent the psychological effect of an inhibition of neural activity in the visual system.

A different form of representation, as in Fig. 54, will show more clearly the relations between stimulus and sensory magnitudes. In the upper part of this figure the stimulus is shown as first of constant magnitude, then suddenly rising and then finally maintaining a uniform level. The sensory effects follow a different course. It is as though a quantity of sensation were displaced from one area (the concave one) to another area (convex).

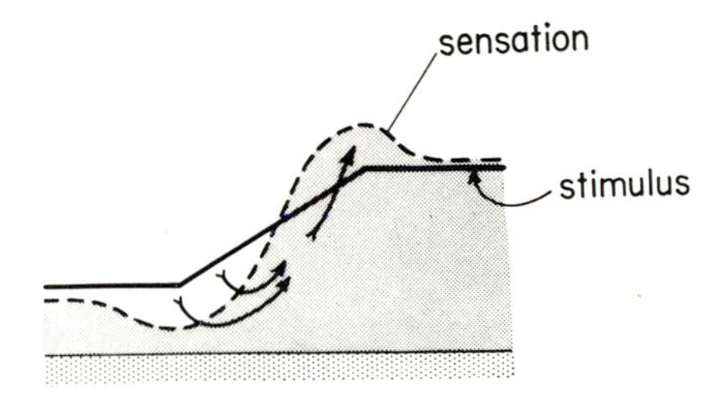

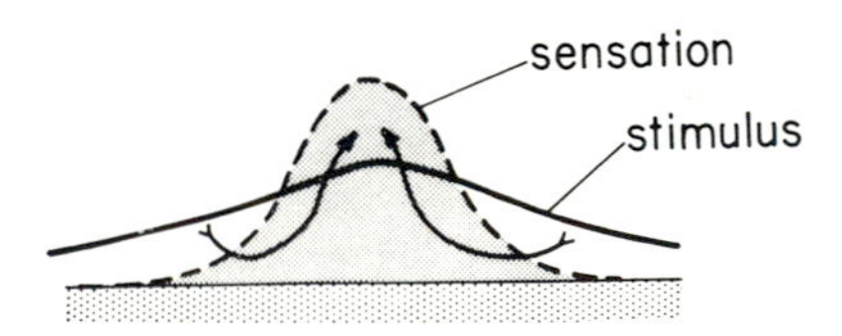

Fig. 54. Further studies of the effects of stimulus discontinuity. Note the presence of undershooting and overshooting in the sensory magnitude. From *Journal of the Acoustical Society of America* [83].

The lower drawing of this figure is a clearer indication of what is going on to modify the local distribution of sensory magnitude. An effect like that pictured can be produced with a rotating cylinder or with other apparatus to be described. The question that is first raised is whether the formation of the Mach bands is specific to the retina or is a general property of the sensory nervous system.

A simple experiment reveals the existence of this phenomenon for the skin for both direct pressure and vibration. A pattern that followed the form of the solid line in the upper part of Fig. 54 was cut out of cardboard and pressed evenly against the soft part of the upper arm. The results are seen in Fig. 55. The overshoot was well perceived, but the undershoot was often lacking. Persons with thin, soft skin were able to detect it, but in others the lateral shearing forces acting on the skin surface tended to destroy the proper pressure pattern.

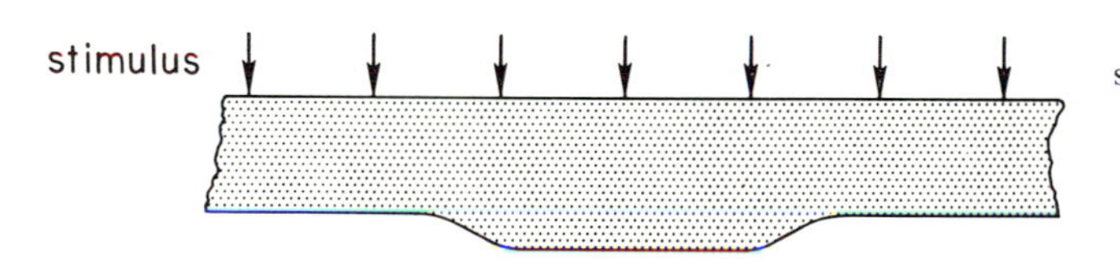

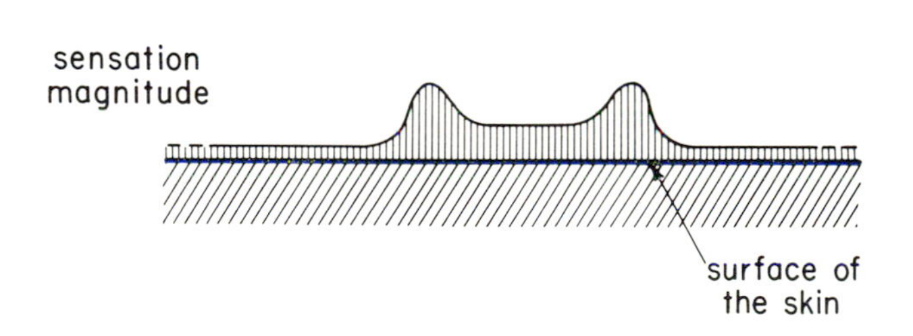

Fig. 55. Effects of stimulating the skin with a contour. Modified from *Physikalische Zeitschrift* [2].

Experiments involving vibrations around 100 cps proved much easier. An arrangement is shown in Fig. 56 by which vibrations can be produced that are constant for a distance, then increase in amplitude, and finally remain at a high amplitude, as the shaded area indicates. When this vibratory pattern was placed on the upper arm, the overshoot was easily located (at *b* in the figure), yet at the same time no vibration was sensed at *a* in the concave region of the stimulation pattern. This was true even when the amplitude at *a* was very strong, as could be verified by stroboscopic examination.

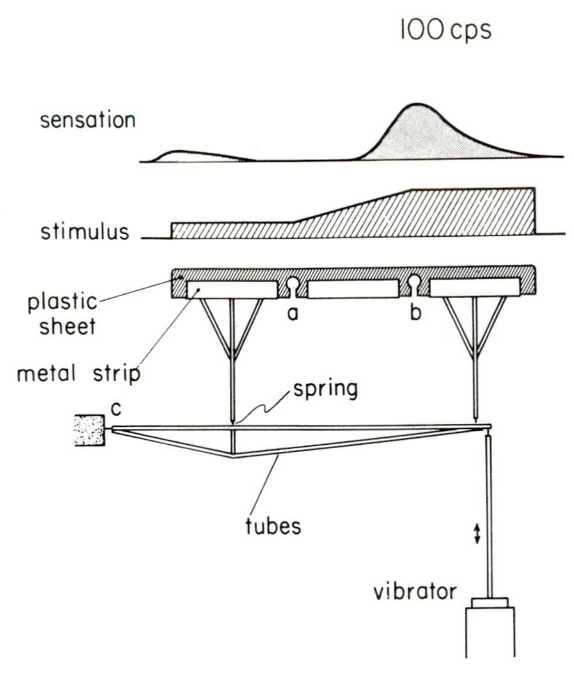

Fig. 56. Effects of vibratory stimulation with a pattern containing discontinuities. From *Journal of the Acoustical Society of America* [83].

Moreover, the inhibition in this region suppressed other effects such as muscle twitches, blood flow, and the spontaneous activity of neural elements.

Under proper conditions this situation produces a unique sensation: the person not only fails to perceive the vibration in the inhibited region, but does not even feel the skin; he has the impression that nothing is there. The experience is entirely different from that ordinarily found in the absence of stimulation.

After success with this experiment, the question arose whether the Mach bands could be accounted for by the scheme used earlier to represent an area of sensation surrounded by an inhibited area. The scheme shown at *a* of Fig. 57 was simplified to produce *b*, a form with only

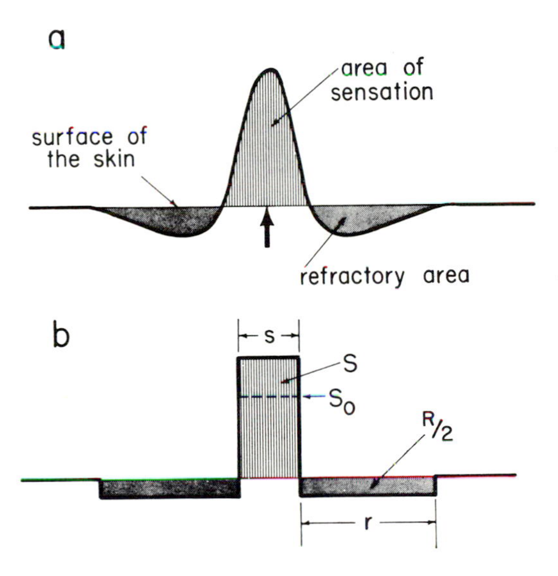

Fig. 57. In *a* is shown the assumed pattern of neural activity for a stimulating point, consisting of a central sensation surrounded by an area of inhibition. In *b* is a simplified form of this pattern. Here *S* represents the sensory area and *R* the total inhibitory area. The distance *s* represents the width of the sensory area and *r* the width of the inhibitory area on each side. The area up to the broken line represents a sensory area S_0 corresponding to one level of stimulation, and the larger area *S* corresponds to a higher level of stimulation. From *Journal of the Acoustical Society of America* [90].

square shapes. With this simplified scheme it was possible to show that two inhibitory units operating in this way would produce a summational effect as shown in *b* of Fig. 58 when the separation between the units was small. This effect continues until a distance of separation is reached, as shown in *d* of this figure, where the summation suddenly gives way to inhibition.

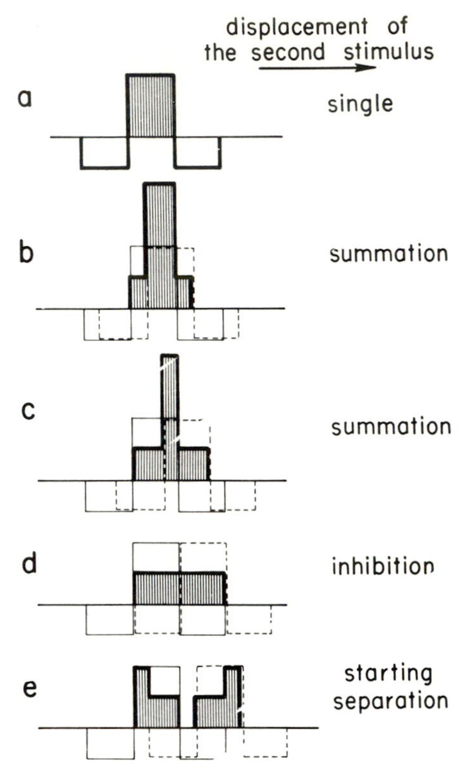

Fig. 58. Effects of combining two neural units of the type described in the preceding figure, when these units are elicited at different separations. The thin solid line represents one unit and the thin broken line represents another. From *Journal of the Acoustical Society of America* [87].

At the same time the total sensory magnitude is less than before. This inhibition enters just before the two sensory maxima fall apart as represented in *e*. It is further possible to deduce that for a stimulus pattern in which the magnitude increases steadily, as in Fig. 59, the sensory pattern will have a regular form without any funneling. This fact is shown by drawing in the sensory effect for each small local stimulus magnitude by itself and finally adding the positive and negative (inhibitory) areas.

On the other hand, if we carry out this same procedure with a pattern like that in the upper part of Fig. 60, the addition of sensitive and inhibitory areas gives the shaded area below, where we find an undershoot and an overshoot just as in the Mach bands. The distortions seen in the forms of the undershoot and overshoot depend largely on the type of inhibitory unit. The larger the inhibitory area the larger the overshoot and undershoot become. From this model it is clear that it should be possible to predict the width of the inhibitory areas from observations on the Mach bands.

Before proceeding with this prediction, however, I wish to discuss the question whether the Mach bands are formed in the periphery, in a

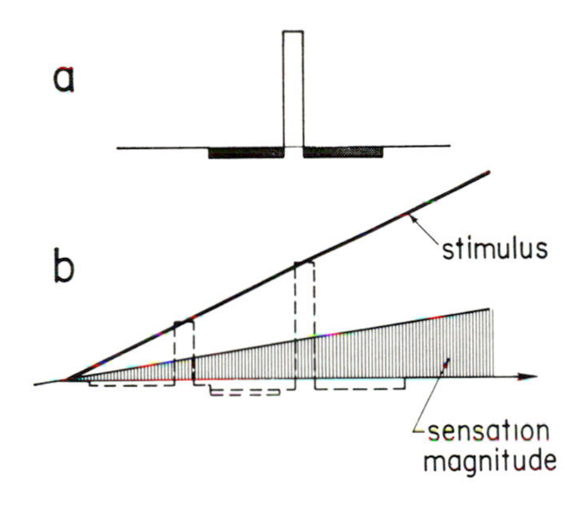

Fig. 59. Explanation of the absence of Mach bands when the stimulation is without discontinuity. From *Journal of the Acoustical Society of America* [90].

single process of lateral inhibition, or perhaps in several steps. The last seems physiologically the most likely.

The geometrical treatment described above could be repeated three times with units a third of normal size, and the results combined or could be done only once with the normal unit, with results that are almost exactly the same (see Fig. 61). This observation means that inhibition occurring in three separate steps can give the same Mach bands as could be produced in a single step.

The mathematics of the Mach bands

Now that the forms of the Mach bands have been shown to result from the action of inhibitory units, some information is required about their dimensions. A method that is suitable for visual observations is the following.

Mach used a rotating disk to produce a light stimulus with a particular intensity gradient.

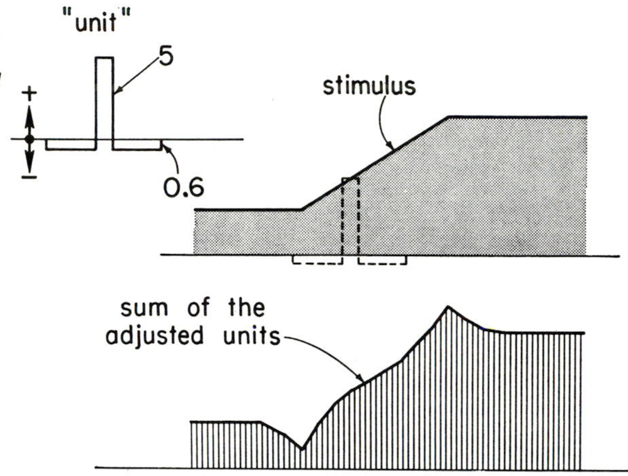

Fig. 60. Effects of the summation of neural units in the presence of discontinuities. From *Journal of the Acoustical Society of America* [87].

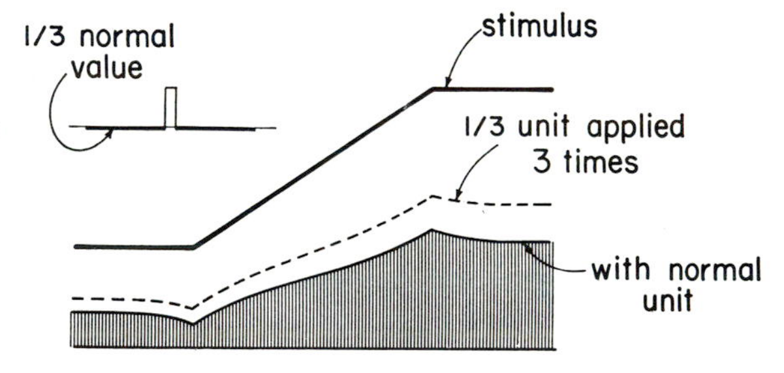

Fig. 61. The repeated action of small neural units compared with the less frequent action of large units. From *Journal of the Acoustical Society of America* [90].

This method has limitations if various patterns are needed or it is desired to change the patterns quickly. Two other methods are more flexible. In one a cylindrical lens is placed above a strip of white paper to "stretch out" the surface in one direction to give parallel stripes whose light density corresponds to the width of the paper strip. In another method a rotating mirror or prism is used to give the same effect. My preference is for the rotating prism. Figure 62 shows the arrangement. Transmitted light was used to enhance the intensity difference between white and black. Most of the stops consisted of double-edged razor blades placed on the diffusing glass of the light box. The blades could be moved to give the opening any shape. A rotating glass prism of square cross section above the

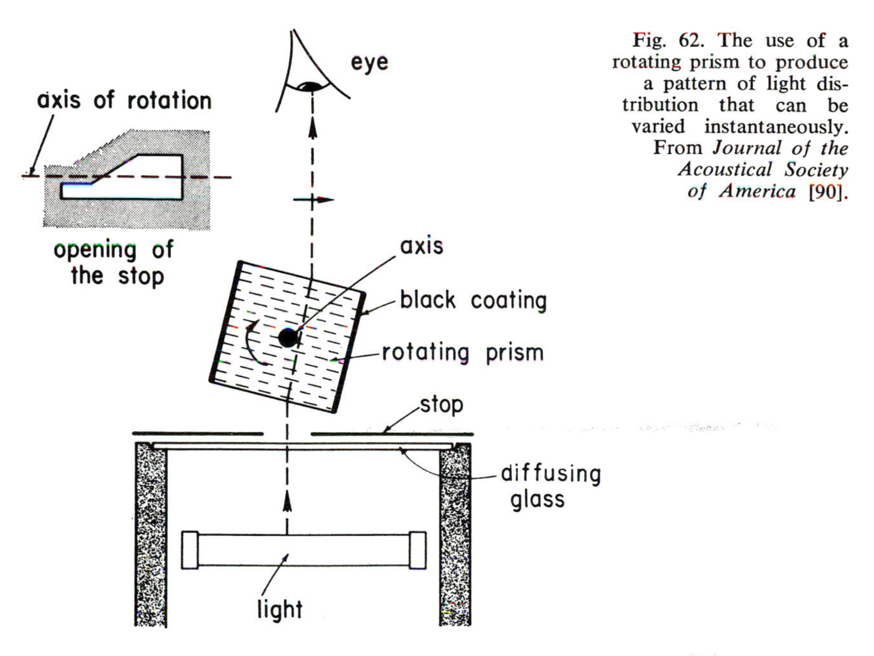

Fig. 62. The use of a rotating prism to produce a pattern of light distribution that can be varied instantaneously. From *Journal of the Acoustical Society of America* [90].

opening transformed the geometrical pattern of the opening into a pattern of light density. To assist in centering the prism on the axis of rotation two opposite faces were coated with black. Therefore the rotation of the prism did not deflect any light beam from its normal direction but only displaced the beams parallel to themselves, in the direction perpendicular to the axis of rotation. The effect was the same as moving the opening at a constant speed perpendicular to the axis of rotation. The prism was rotated with a synchronous motor attached to a low-frequency oscillator, which made the speed of rotation easy to adjust. Covering the edge of the opening with black metal sheets whose edges were differently shaped gave a variety of Mach bands.

To define the shape of the neural unit in vision, it is necessary to determine the width s of the sensory area, the width r of the refractory (or inhibitory) area on one side, and the ratio between the sensory area S and the total refractory area R (on both sides) which is S/R. (See Fig. 57.)

It is relatively easy to find the width r of the refractory area, for the total extension of the Mach bands is twice this width. This relation can be seen immediately by comparing b in Fig. 63 with c.

A better method for measuring the width r of the refractory area is shown in Fig. 64. The visual stimulus is a trapezoid whose sloping sides have an angle of 45°. As the length d of the flat top is increased from $r/2$ to $2r$, it will be seen that at $r/2$ the sensory pattern is flat, and the observer sees only a single white band. But for the other values of d there are two sep-

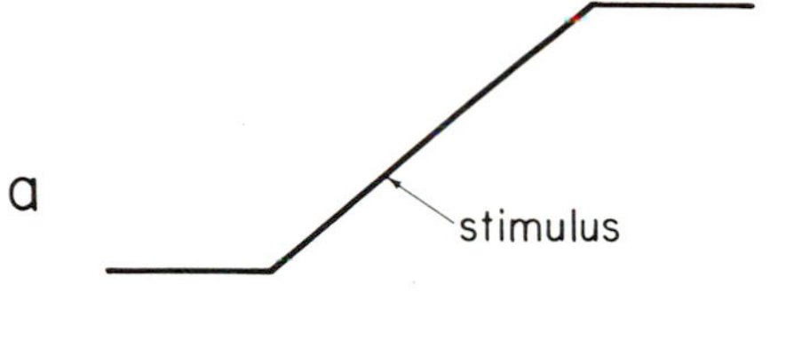

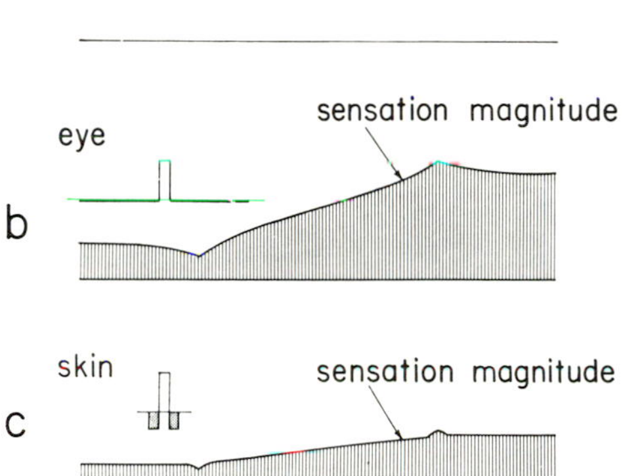

Fig. 63. The widths of the Mach bands can be accounted for by the widths of the inhibitory areas. From *Journal of the Acoustical Society of America* [90].

arate bands at the edges. This was the criterion used to determine r. The distribution of the sensory magnitude along the opening in Fig. 64 was calculated by the method described earlier and by use of the eye unit shown, for the different trapezoids.

The observations were made by placing two razor blades at an angle of 90° on the glass plate, as at the top left of Fig. 64. The base of the triangle was formed by a third blade, which was parallel to the rotating prism. The peak of the triangle could be cut off by moving a fourth blade downward by means of a micrometer screw until a darker band began to appear in the middle. The light intensity in the opening was 100 millilamberts and it was presented to the eye 60 times per second. The critical width of the top of the trapezoid was 0.45 mm, viewed from a distance of 25 cm. Thus $r = 0.9$ mm for an object seen from a distance of 25 cm. The value of r depends on the light intensity, and if the stimulus is interrupted it depends also on the number of presentations per second.

It is more difficult to determine the value of s, the width of the sensation area. This width is small compared with r, because it represents the smallest distance between two points that can be separated by the eye. Fortunately the exact value of s is not important as long as $s \ll r$. It can be seen from Fig. 65 that the distribution of the perceived magnitude does not change when s is reduced to half but the sensory area S remains the same.

The brightness difference between the Mach bands and the area surrounding them is determined mainly by the ratio between the perceived area and refractory area. One way of determining S/R is to select a stimulus distribution that will give inhibition over its entire extent. The stimulus distribution is then varied until one is found in which the distribution of sensory magnitude is constant along the entire extent. For example, if we drill a precisely round hole in a thin metal sheet and observe it through a rotating prism the distribution of brightness across the light bands is as shown in *a* of Fig. 66. If, however, we cover the lower part of the round opening with a straight edge parallel to the axis of rotation of the prism, as in *b*, and move it up and down over the opening, it is possible to obtain a distribution of light that is constant except at the edges. This pattern was found to be most nearly flat when the straight

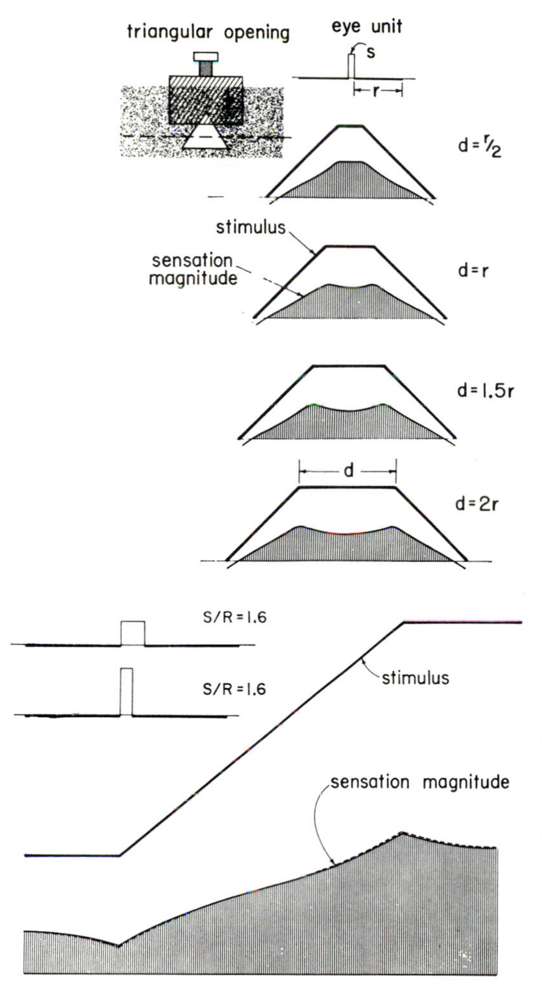

Fig. 64. Method of determining the width of the refractory area. From *Journal of the Acoustical Society of America* [90].

Fig. 65. The size *S* of the sensory area determines the perceived pattern, regardless of the width of this area, as shown (up to the point where this width exceeds the height). Dotted line corresponds to the upper form of inhibitory unit. From *Journal of the Acoustical Society of America* [90].

edge covered one-fourth of the diameter of the opening. As the straight edge was moved farther toward the center the pattern again became rounded at the top as in *a*. The same results were obtained with openings varying in diameter from 2 to 5 mm.

Fig. 66. Method of modifying the light distribution by moving a straight edge across a round light opening. In *b* a setting is shown that produces a uniform magnitude of sensation across the opening.

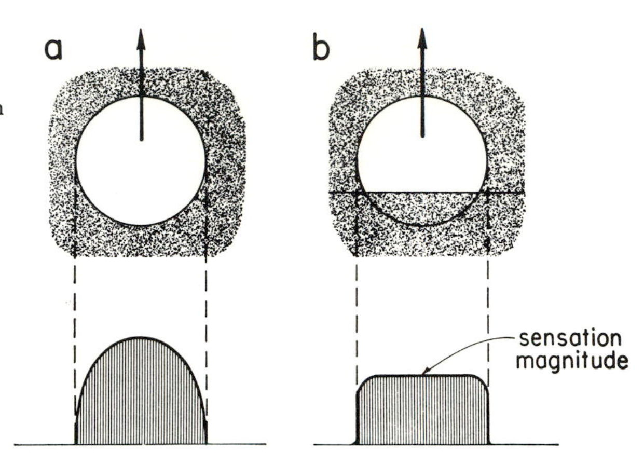

To determine S/R from this observation, the distribution of light intensity produced by the opening in *b* of Fig. 66 was plotted in *a* of Fig. 67. An arbitrary unit was taken, with *s* about one-eighth of *r*. The area of *R* was varied until a relatively flat pattern of perceived magnitude was obtained, with S/R equal to 1.6. As may be seen, if S/R is made equal to 1.9 or 1.3, the pattern becomes less flat. The drawing *b* of Fig. 67 shows that this situation is changed when the diameter of the opening is increased threefold. Under these conditions an S/R of 1.2 gives the most nearly flat distribution of sensory magnitude. In these observations the diameter of the opening was 5 mm, which is about 6 times the width of the visual refractory area.

These observations for vision are summed up as follows. The value of *r* was found to be equal to 0.9 mm for an object seen at a distance of 25 cm, and S/R is equal to 1.6 (see Fig. 68*a*). These quantities vary with the light intensity, and if the light is interrupted they vary with the interruption rate.

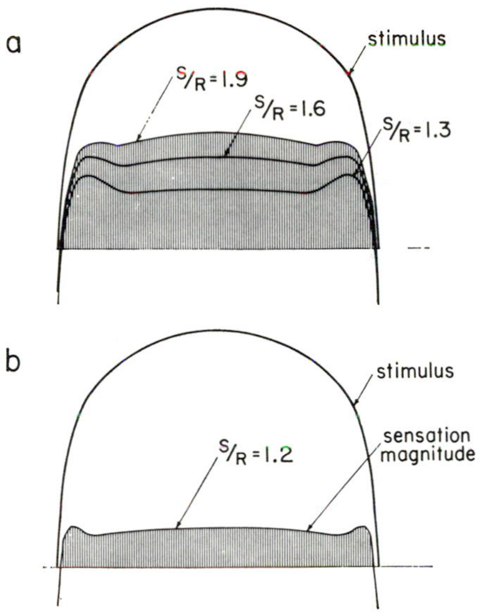

Fig. 67. Effects on the form of the perceived pattern of varying the ratio of sensory to inhibitory areas S/R. From *Journal of the Acoustial Society of America* [90].

Similar observations were made on the skin of the lower arm by applying cardboard outlines of different forms. From the results a determination was made of the dimensions of the lateral inhibitory area and the perceived area. The results are given in the lower portion of Fig. 68.

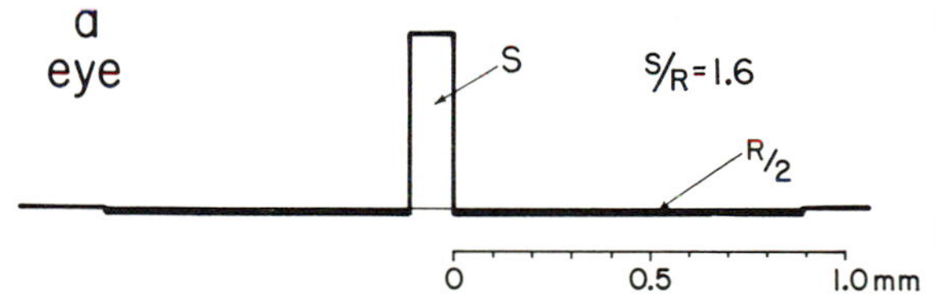

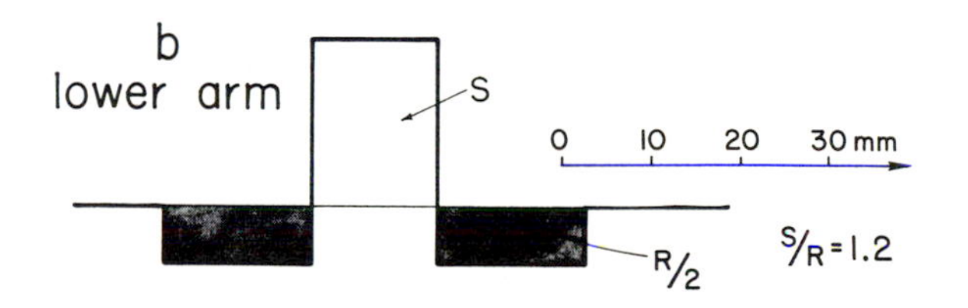

Fig. 68. The dimensions of the sensory unit as determined for the eye, *a*, and for the lower arm, *b*. In *a* the visual field was viewed at a distance of 25 cm. From *Journal of the Acoustical Society of America* [90].

Fig. 69. When the width s of the sensory unit is about the same as the width r of the inhibitory area a minimum of sensory magnitude is obtained when two sensory units are adjacent, as at *a*. When the two sensory units are farther apart, as in *b* and *c*, the inhibitory effect declines and the sensory magnitude is the same as for single-point stimulation. From *Journal of the Acoustical Society of America* [90].

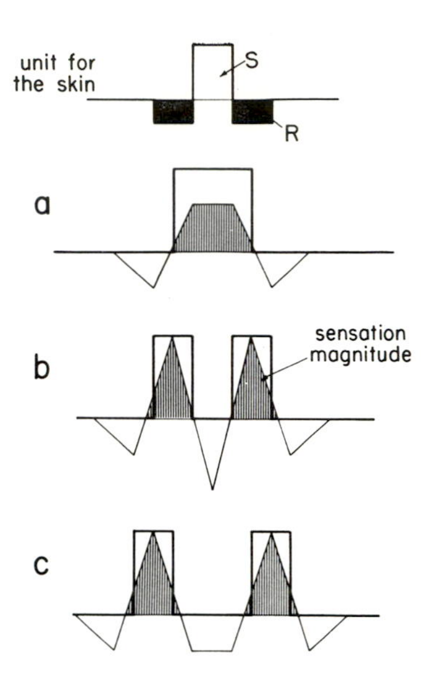

There is an interesting difference between the two-point thresholds of the eye and the skin. For the skin, as shown in Fig. 69, the magnitude of the sensation (shaded area) is a minimum when the refractory area has about the same width as the sensory area. This happens just before the sensation begins to separate into two parts (shown at *a* in the figure). On the other hand, when as in vision the neural unit has a wide inhibitory area, there is little change in the peak of the sensory magnitude when the separation of the stimuli is increased; this situation for the eye is shown in Fig. 70. Here the brightness of the two points does not decrease as they are brought close to one another. It should be pointed out, however, that tremor of the eye may smooth out the perceived magnitudes.

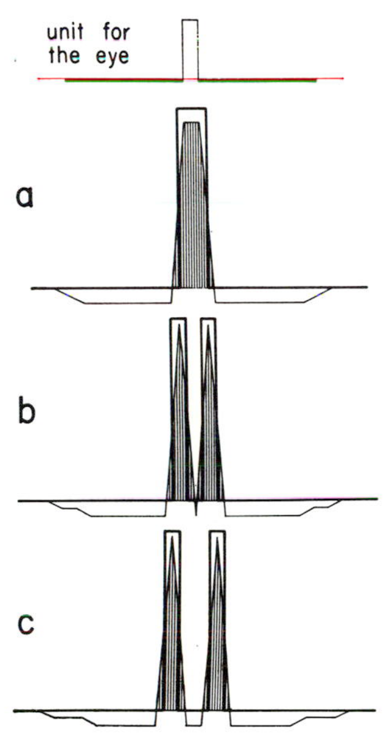

Fig. 70. For the eye, stimulation at two adjacent points fails to show the minimum as found for the skin. This difference can be explained by the assumption that the inhibitory area for the eye is relatively large. From *Journal of the Acoustical Society of America* [90].

The use of geometric models of summation and inhibition in the lateral sensory processes as just illustrated should make it possible to discover what kind of process determines the formation of the Mach bands and similar phenomena. For example, Fig. 71 shows how five neural units may be combined to form a new unit that represents an inhibitory area in which there is a progressive decrease in inhibition from the center of the stimulus outward. In other respects this pattern is similar to the single units from which it was constructed, and it will produce Mach bands in the usual way. From this evidence we may conclude that a gradual decrease in inhibition with distance from the stimulus is not an essential feature of inhibitory areas.

Fig. 71. The effects of combining five neural units in a continuous line.

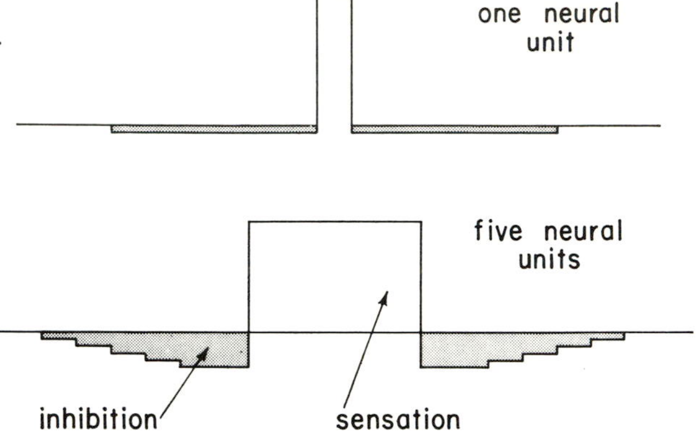

Similar changes may be considered for the neural unit. Figure 72 represents two situations that are alike except that in the one shown in the lower part of the figure the inhibitory areas have been displaced laterally away from the point of stimulation. This displacement increases the overshoot and the undershoot and also the width of the Mach bands. This effect is of interest from a physiological standpoint because it shows inhibition away from the stimulated region to be more active than that within the region itself. It seems proper to assume that in the nervous system the width r of the inhibitory area may increase laterally as stimulus magnitude increases, as suggested in the upper part of Fig. 73. Such an increase in the lateral spread of inhibition will widen the Mach bands for the convex form of amplitude distribution of the stimulus at the expense of the concave form. In other words, the "white" bands become larger relative to the "dark" bands. When the inhibition remains constant and is independent of stimulus magnitude, there are no Mach bands, as the lower drawing of this figure shows.

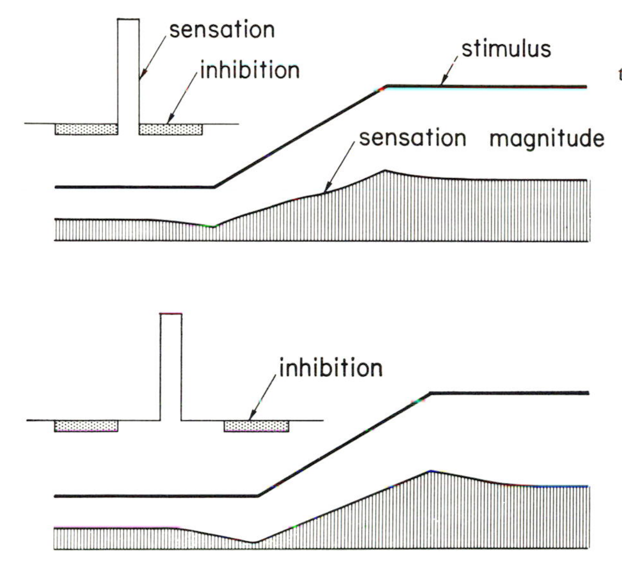

Fig. 72. The effects of varying the location of the inhibitory area on the width of the Mach bands.

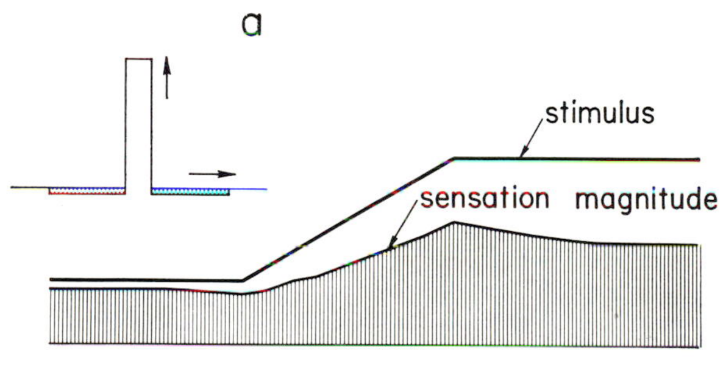

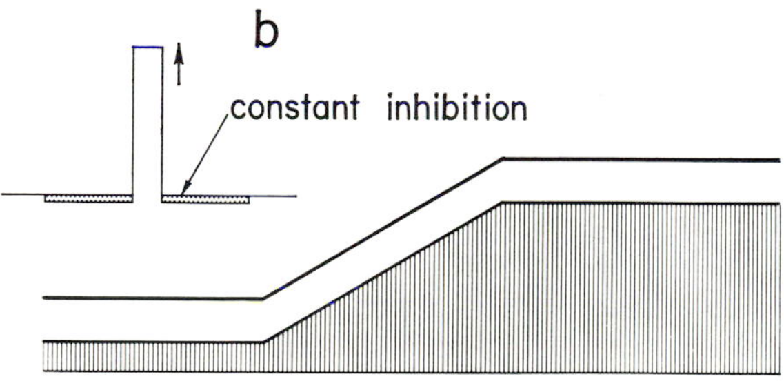

Fig. 73. When the width of the inhibitory area increases along with the increase in the stimulus (and in the sensation), the Mach bands appear as in *a*; but when the inhibitory area remains constant and the stimulus increases, there are no Mach bands, as in *b*.

This evidence seems to indicate that the production of Mach bands requires a change of inhibitory pattern as a function of stimulus magnitude. Yet if this change of inhibitory pattern is inversely proportional to the stimulus intensity rather than directly proportional to it, again there are no Mach bands. In the upper part of Fig. 74 the area of inhibition was decreased as the stimulus intensity was increased. In the lower part of this figure the inhibitory area was kept constant as the stimulus intensity was changed, but the magnitude of the inhibition was decreased along with the increase in stimulus intensity.

From these considerations we might conclude that a neural unit of the type shown in Fig. 68 comes close to a proper explanation of the Mach bands. Yet it should be mentioned in this connection that the shape and magnitude of the sensory unit depend on the stimulus magnitude. The values indicated are for a middle range of stimulus intensities. As shown earlier in Fig. 27, there is no inhibition but only summation in the region of the threshold.

The simple fact that a sensory area is surrounded by an inhibitory area seems to be sufficient to account for the Mach bands and similar phenomena of lateral inhibition. The shape of the lateral inhibitory area does not seem to be of primary significance. This conclusion can be reached also in consideration of six different graphical and mathematical models of inhibitory networks presented by Ratliff (1965, p. 122), which perform similar functions.

In Fig. 75 is shown the relation between perceived patterns of intensity and a stimulus mag-

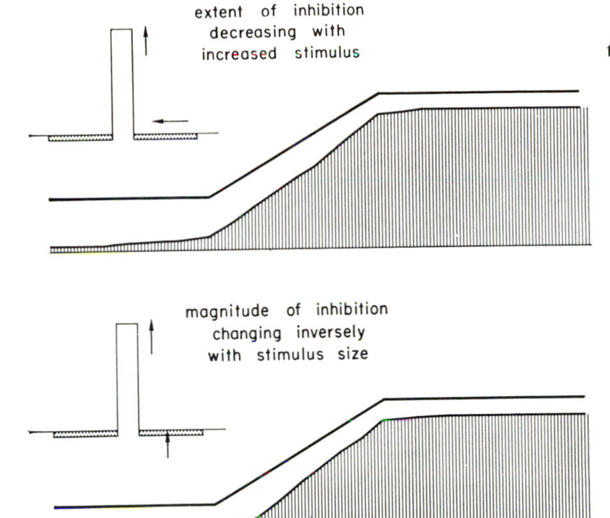

Fig. 74. The absence of Mach bands when the width of the inhibitory area decreases, or the magnitude of the inhibitory effect decreases, along with an increase in stimulation.

nitude that varies in space sinusoidally. The sensation also has a sinusoidal form, with a slight increase in peak amplitude only when a wavelength approximates the width of the inhibitory area.

A spatial pattern of stimulus intensity that is easy to observe, and also is probably the most important physiologically, is a sudden step change of intensity. The distribution of sensory magnitude produced by such a step is shown in *a* of Fig. 76 for a neural unit of the normal type. Here we find the usual undershoot and overshoot. In *b* the sensory magnitude was plotted for a unit with a laterally decreasing inhibitory area, produced by an increasing stimulus magnitude. Here we find no undershoot or overshoot.

Fig. 75. Application of the neural unit concept to sinusoidal stimulation, with repetition patterns varying in their spacing.

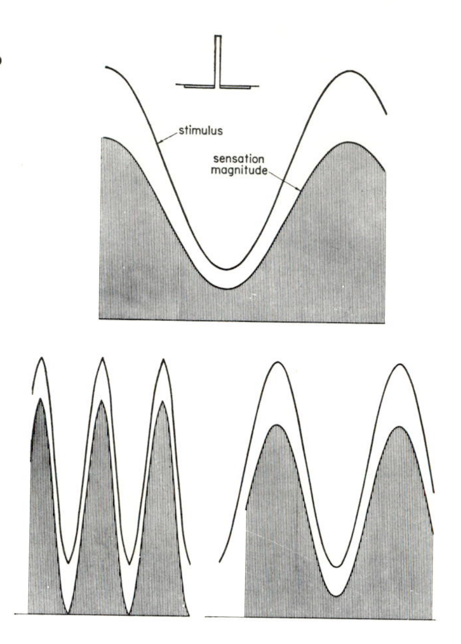

Fig. 76. The neural unit concept applied to a step function. In *a* the normal unit is used (in which the S/R ratio is fixed), and in *b* a unit is used in which the width of the inhibitory area decreases with an increase in stimulus magnitude.

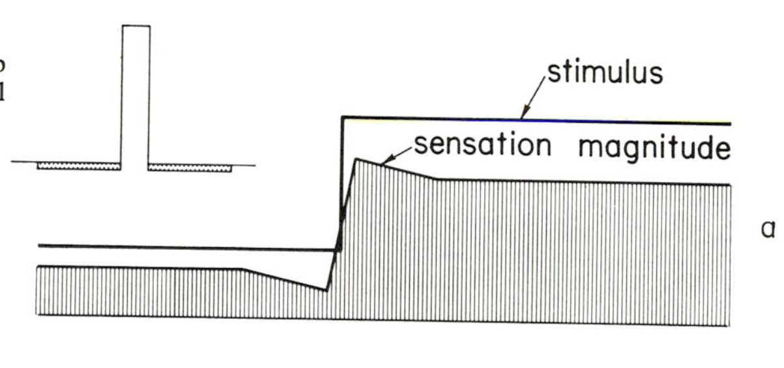

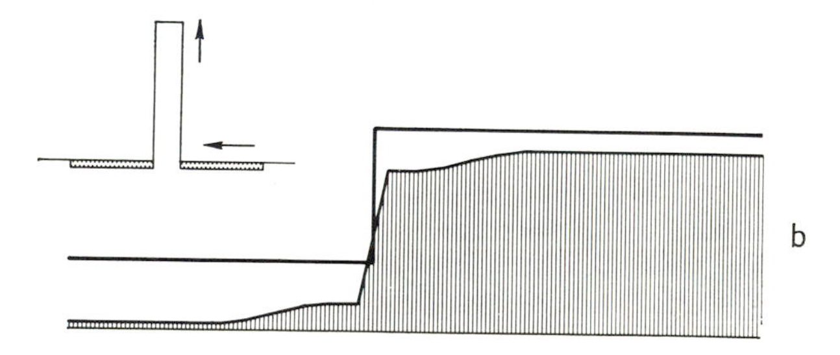

In a measure we can select among the various possibilities of neural interconnections the ones that can be used to account for the Mach bands. Probably the model of the normal neural unit as shown in *a* of Fig. 77 ought to be replaced by that shown in *b* of this figure; here there is inhibition over the entire surface and a superimposed sensory pattern. Parts *a* and *b* of this figure are identical from the point of view of the geometrical procedures previously used, but I believe *b* to be the more probable physiologically.

The possible neural interconnections are shown in *c* and *d* of Fig. 77. It is possible that a single stimulated end organ transmits the

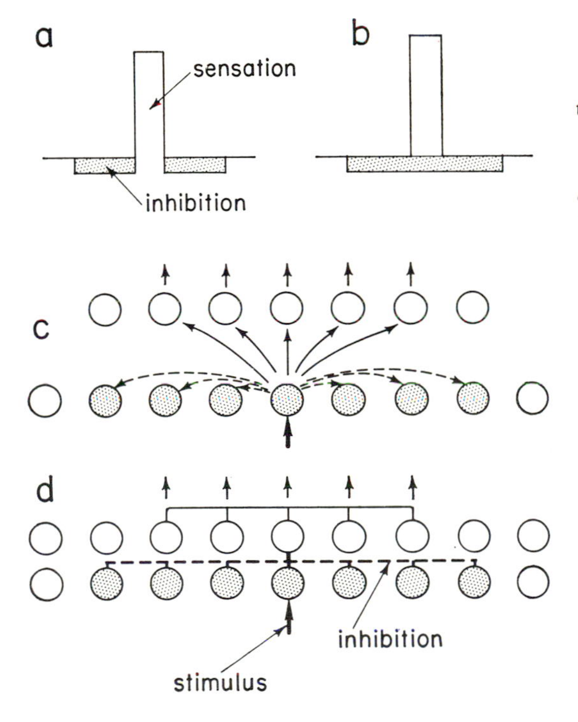

Fig. 77. In *a* is shown a schematization of a neural unit of the form usually employed in the foregoing discussion, and at *b* another form that could be used equally well. In *c* and *d* are shown equivalent forms of interconnection among the sensory and neural cells in the region of the stimulus. Shading represents inhibition.

stimulus to several nerve cells at a higher level, and at the same time produces an inhibition of neighboring end organs. It is also possible that a single stimulated end organ transmits the excitation to a higher level, and that there is a lateral spread that produces sensory effects. At the same time the lateral sense organs are largely inhibited as indicated by the shading. The extent of the inhibited area is the same as that shown in *b* of this figure. The extent of the area producing the sensation is given by the lateral extent of the cells transmitting the effects. The only requirement for the production of Mach bands and similar phenomena is that the inhibited area be larger than the transmitting area.

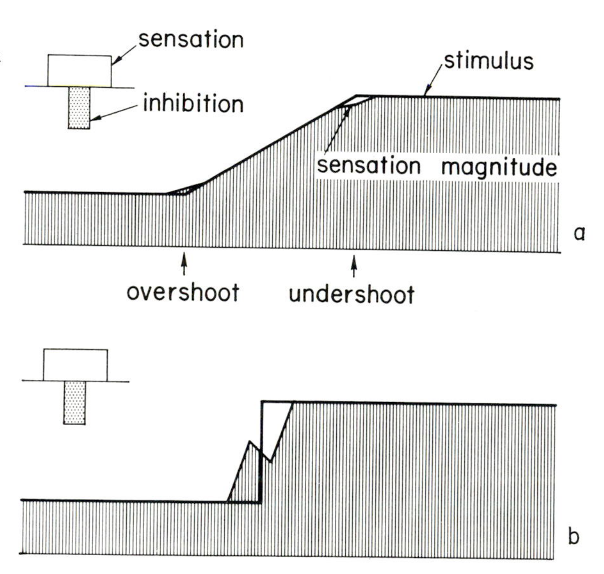

Fig. 78. The effects of making the inhibitory area smaller than the sensory area, for a ramp function *a*, and for a step function *b*.

To confirm this relation we can study the distribution of sensation produced by a neural unit, as indicated in Fig. 78, in which the inhibitory area is smaller than the stimulated area. As may be seen, the undershoot and overshoot are reversed as compared with the observations made in vision and skin sensitivity. Clearly the stimulated area ought to be smaller than the inhibitory area.

At the moment it is possible that the drawings of either *c* or *d* of Fig. 77 represent the true situation. It is even possible that the two layers of cells represented in these drawings combine into one, and that there is but a single layer of nerve cells containing some cells that are stimulated and partially inhibited at the same time.

Fig. 79. The effect on localization of the time delay between two stimuli: *A*, on the skin; *B*, at the two ears. From *Journal of the Acoustical Society of America* [79].

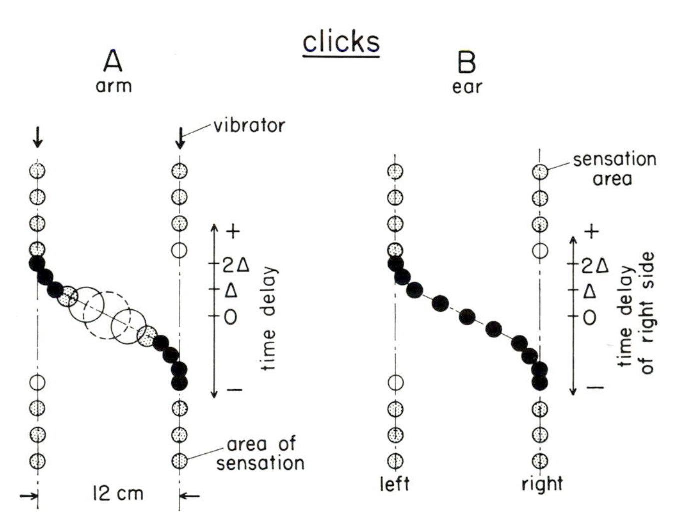

III · INHIBITION AS A RESULT OF TIME DELAY

Interaction between two stimuli with a time delay

When there is a time delay between two stimuli, inhibition plays an important role. An example is directional hearing, first described in detail by von Hornbostel and Wertheimer (1920). It was found possible to duplicate the phenomena on the skin (Katz, 1937). In Fig. 79 results are shown for both the ear and the skin. For the skin two electrodynamically driven vibrators were used, which touched the skin with a point 3 mm in diameter. Care was taken that this vibrator when loaded by its contact with the skin produced a displacement proportional to the current through the electrodynamic unit. For this purpose it was necessary to record the vibrations for different frequencies and to make sure that the transients produced by the combined system (skin and driving unit) were very brief. The best testing method was to "pluck" the electrodynamic system by switching a direct current on and off while observing the result with an electrostatic pickup unit in contact with the skin. It was also essential to have two driving units as alike as possible. I find it convenient to have the natural (resonance) frequency of the whole mechanical system very high, and also to have a high mechanical impedance looking into the vibrator. These conditions require the use of stiff springs in the suspension of the driving unit. The result of these conditions is that the units are independent of the mechanical impedance of the skin where the tips are applied. Because it turned out that phase distortions are easily detected by the nervous system, it is necessary not only to know the nature of the transients produced by frequency distortion but also the nature of those produced by phase distortion.

If two vibrators are placed 12 cm apart on the arm, and are actuated with a series of clicks without time difference, the vibration will be localized in the middle between the two vibrators as a rather diffuse sensation. This effect is represented in *A* of Fig. 79 as a large broken circle. When a time difference between the vibrators is introduced, the sensation will move toward the vibrator that receives the click earlier. At the same time the lateral spreading of the vibratory sensation becomes smaller. When the time difference is as large as one or more milliseconds the whole sensation is localized beneath the tip of the vibrator that is actuated first, and the second vibrator seems to produce no sensation at all; it is almost completely inhibited. If we reverse the time relation so that the second vibrator receives the clicks earlier, the vibratory sensation will shift to the other side. After a few days of training these shifts are easy to observe, but it is important to ensure that the perceived magnitude produced by the vibrators is exactly the same. If this is not the case the regular form of the localization pattern shown in Fig. 79 is modified and becomes asymmetrical.

The inhibition resulting from a time delay of one vibrator of one or more milliseconds does not have the complete form of the inhibition seen in antagonistic muscles. It is more a funneling of the localization from one side to the other, as shown by the fact that switching off one of the vibrators produces a decrease in the magnitude of the sensation.

The drawing in *B* of Fig. 79 shows a similar situation for the ear when stimulated by a series of clicks. Here again care was taken to make the loudness of the click series in right and left ears exactly the same, which was done by adjusting the input when the clicks were presented alternately.

A difference between auditory and vibratory sensation is that the acoustic image does not change in size as it is caused to move from the middle to one side or the other. This fact is an interesting one, and indicates that the funneling processes involving the two ears are somehow different from the funneling process on the skin. Yet the time delays necessary to shift the sensation from the middle all the way to one side are the same in the two cases.

A peculiar phenomenon in directional hearing is the sensation of a rotating tone. It can be observed by presenting to one ear a pure tone of 500 cps and to the other ear a tone of 500.5 cps. If the tones are adjusted for equal loudness and then turned on simultaneously we hear only one tone of a certain pitch, but this tone changes its position in space at a rate determined by the frequency difference in the two ears. In this example the period of rotation is 1/0.5 or 2 seconds.

In general this tone is perceived as moving around the head, hence the term "rotating tone." Some persons have difficulty in observing it at all, and others have difficulty in following the position of the sound image.

The rotating tone was the subject of much discussion a few decades ago, as it was not clear whether it was the result of the changing phase differences during the cycle or was an effect of the time delay. It finally turned out that it is the time delay that determines the localization. At the instant the phase difference is zero the sound image is in the median plane, just as for a series of clicks, and as the phase becomes different in the two ears the image moves to the right or left.

The usual analogous relation between auditory and vibratory sensations led to the attempt to produce a rotating effect on the skin. For this purpose two vibrators were applied to a fingertip, one on each side as shown in Fig. 80, and adjusted to give equal sensations when presented alternately. When they were turned on simultaneously the observer felt a continuous movement on the finger from left to right and back again just as with the rotating tone. The main difference is that on the fingertip there is during the movement a change in the lateral spread of the sensation and usually in its magnitude also, as illustrated in Fig. 80. The magnitude is least when the sensation shows its maximum spread.

These analogous relations between hearing and vibratory sensation raised the question whether they were determined by a general character of the nervous system. If so, the same

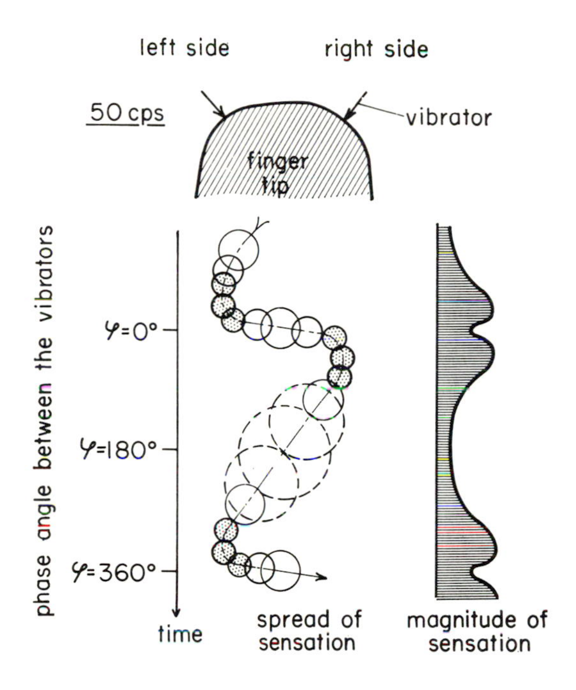

Fig. 80. The effect, on localization, of the phase relation between two vibrations on the fingertip. From *Journal of the Acoustical Society of America* [79].

relations should hold for taste and smell. These sense organs are of special interest because it is generally accepted that they have long reaction times, and if this is true they should require large time differences for a lateral displacement of sensation.

The main difficulty with experiments on taste and smell is in a precise control of onset. The more I attempted to carry out these experiments the more I became convinced that the poor localization found for the sensations was due mainly to poor equipment. A method for dealing with taste is shown in Fig. 81. A steady stream of water was made to flow over a small section of the tongue, and by means of a strong electro-

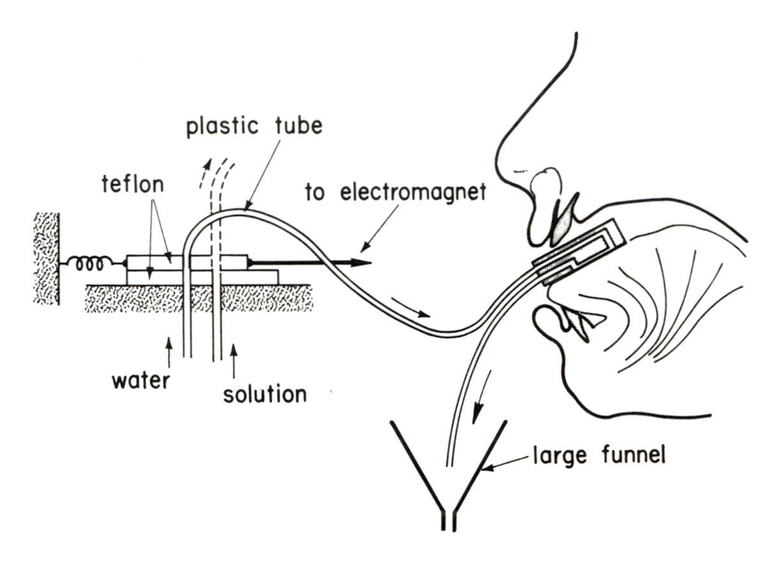

Fig. 81. Method for stimulating the tongue with chemical substances. From *Journal of General Physiology* [111].

dynamic driving unit it was possible to change from water to a test solution of salt, sugar, acid, or quinine. First attempts were made with a plastic plate on the tongue with an opening 1.5 mm wide and 7 mm long, as shown in Fig. 82. It soon became clear that the long tube (Fig. 81) leading from the switching device to the tongue permitted a mixing of the two fluids and destroyed the sharp onset of the change from one fluid to the other. Therefore a new method was devised, as shown in *A* of Fig. 83, in which the switching device was combined with the plate on the tongue. In this device a Teflon slider was shifted along a stainless steel plate by means of a strong electrodynamic driver. The time course of the change in fluids was checked by adding a dye (methylene blue) to one of them and observing the color change through a fiber-optics cable whose end was located close to the point at which the fluid first reached the tongue.

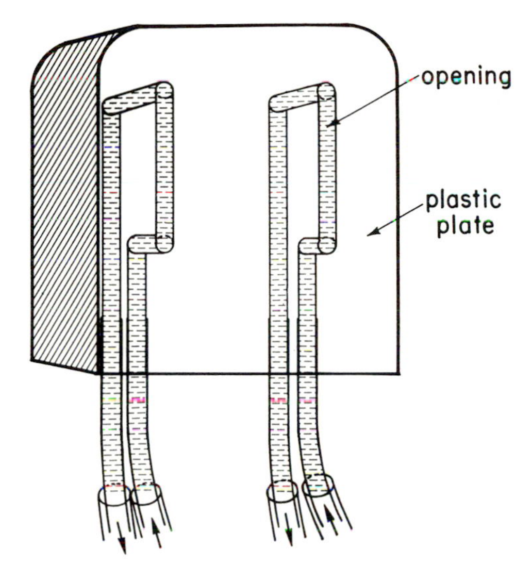

Fig. 82. A plastic plate containing two fluid channels, for taste stimulation. (This figure is the same as Fig. 7.) From *Journal of General Physiology* [116].

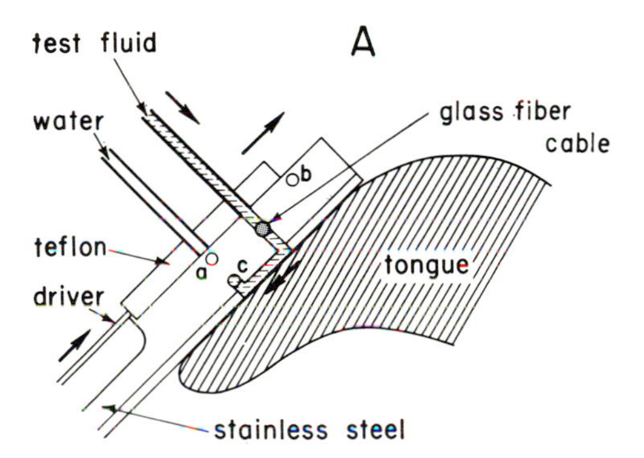

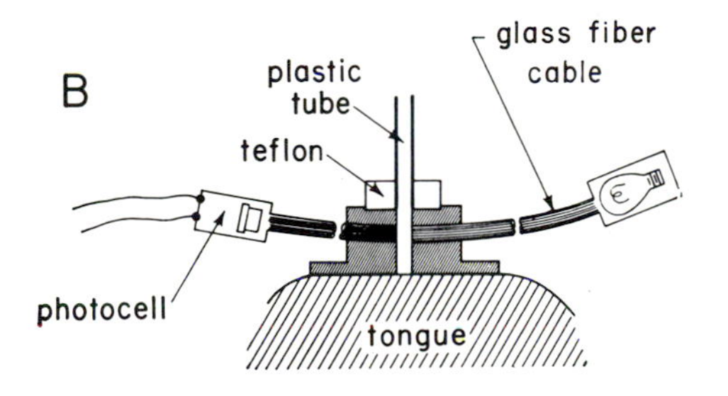

Fig. 83. Method of recording the flow of fluids over an area of the tongue.

A second method of observing the form of the fluid change was to measure the change in electrical resistance in the channel over the tongue. This method was used only with salt and acid solutions, but it had the advantage that the changes were observed at the tongue surface. Special bridge circuits were used for oscillographic recording of the onset of the chemical stimulus. All these developments led to the elaborate setup pictured in Fig. 84.

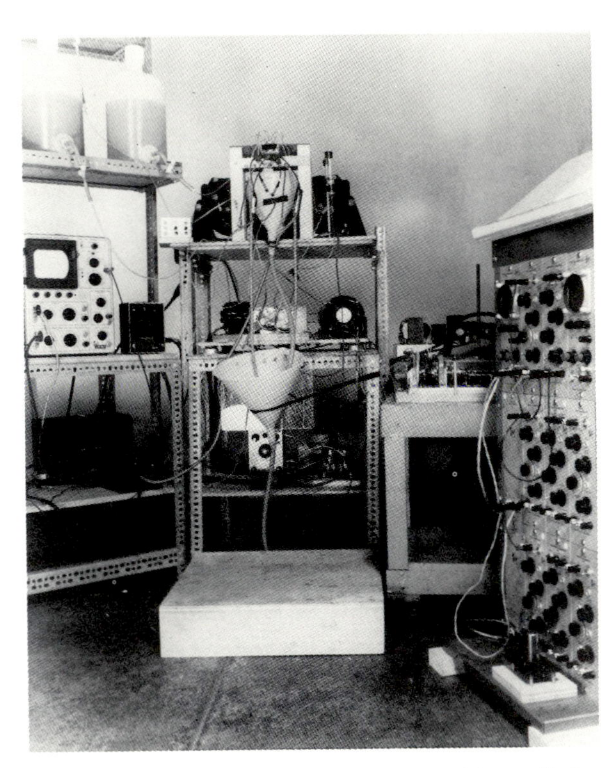

Fig. 84. Equipment used to present two synchronized taste stimuli.

The first experiment was a repetition of the two-point stimulation of the tongue. For this purpose a plate with two openings, each with

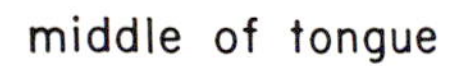

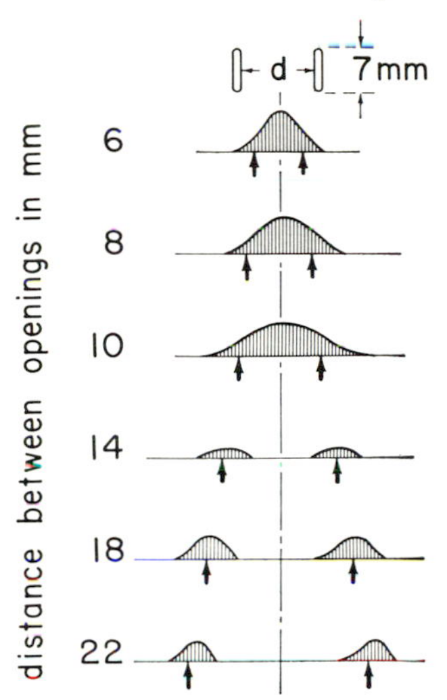

Fig. 85. Effect of the distance of separation of two stimuli on the tongue. Inhibition appears at a distance of 14 mm. From *Journal of General Physiology* [111].

a switching device, was placed on the tongue; the stimulation consisted in simultaneously replacing water with a test solution in each opening. This stimulation lasted for one second. If the distance between the two lines of stimulation was about 6 mm the taste sensation had a maximum that was located on the tongue at a point midway between the two, as indicated in Fig. 85. If the distance was increased there was a lateral spread of the sensation like that for steady pressures or vibratory sensations on the skin. At a certain distance, which was 14 mm in the case represented in Fig. 85, the magnitude of the sensation showed a definite decline. (This magnitude is then smaller than it is for a single stimulation on one side or the other.) As for skin stimulation, the inhibition appears just before the sensation breaks into two separate regions. The distance between the two lines of stimulation at which the inhibition occurs varies considerably in different regions of the tongue, as indicated in Fig. 86. The experiments involving localization were usually carried out in the region where the difference limen was a maximum, because it was easier for the observer to locate the position of the sensation on one side or the other and to follow its movements.

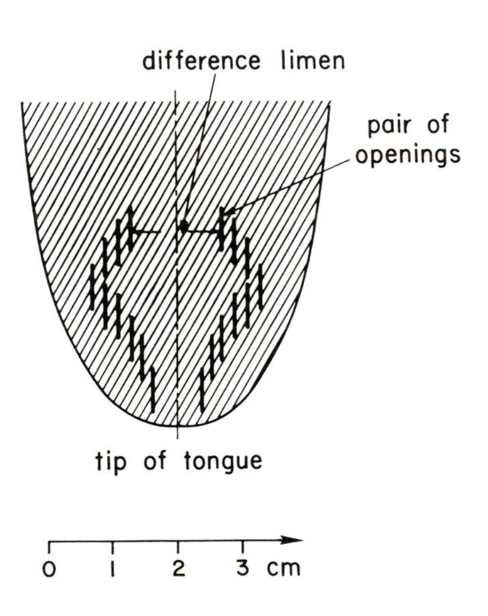

Fig. 86. Variation in the magnitude of the two-point threshold for taste, according to location on the front part of the tongue. From *Journal of General Physiology* [111].

With the apparatus shown in Fig. 83, the concentration of the test solution was first adjusted for equality on the two sides of the tongue, and then the two stimuli were switched on and a check made to ensure that the onset was as short and sharp as possible. There is always an eddy at the front of the new solution, which modifies the onset somewhat. The onset time was a few milliseconds. It was always possible under these conditions to cause the sensation to be localized in the middle of the tongue, though at times it was necessary to introduce a time delay of a few milliseconds in one of the stimuli. The setting thus determined is taken as the "zero position." It was found that the same time difference was required relative to this setting to displace the localization to the right or to the left extreme position. As Fig. 87 shows, this time difference for complete lateral

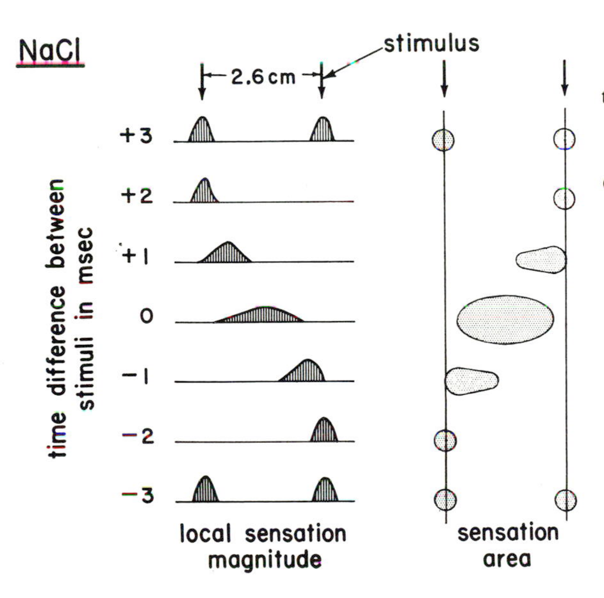

Fig. 87. Localization and extent of taste sensation as a function of the time difference between two stimuli, 2.6 cm apart on the tongue. From *Journal of General Physiology* [111].

displacement was of the order of 2 milliseconds. Again, the action is more a funneling than a simple inhibition. The magnitude of the sensation in the lateral position is greater than it is for a single stimulation.

In other respects there is a close analogy with the phenomena of auditory localization. If the time difference exceeds 3 milliseconds the sensation breaks into two parts. If there is no time difference, there is great lateral spread of the central image. The situation resembles the one with two vibrating stimuli with a large separation between their points of application on the skin.

If there are no time differences between the two stimuli but intensity differences in the two sensations are produced by increasing the concentration of the test solution on one side, there are similar changes in the localization and form of the taste sensation, as shown in Fig. 88.

The brief time differences necessary to shift the localization of the taste sensation to one side or the other is surprising in view of the long reaction times generally thought to exist in this sense organ.

The sense organ of smell is also of special concern because it likewise involves chemical processes with long action times. Also it is clear that many complications can arise from lack of symmetry in the air passages of the nostrils. To provide a clear opening of both passages a nasal spray (0.2 per cent of Otrivin) was used by all the observers, and this procedure seems to be effective in standardizing the aerodynamics of the smelling process.

A second peculiarity of smell experiments is that we perceive odors only during sniffing. Therefore a test situation requires an apparatus that switches in the odor at the proper time in relation to breathing, which is a few tenths of a second after an inhalation begins. This switching can be done by a relay that is triggered by chest movements or it can be handled by the observer himself. The best time relation to the stimulus has to be determined by experiment.

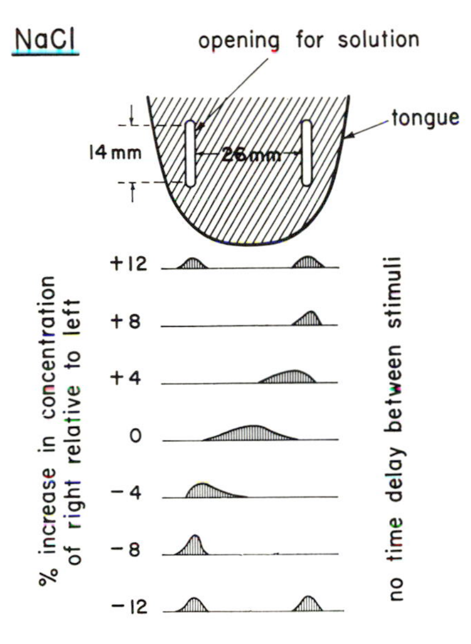

Fig. 88. Localization on the tongue as a function of an imbalance in the intensity of two stimuli. *From Journal of General Physiology* [111].

A complicated set of mixing bottles, temperature regulators, and humidifiers had to be used to make stimulation precise, and a good deal of experience was necessary to avoid errors. It was interesting that as the problems of handling the stimulus were solved and the apparatus was improved in precision, the performance of the observers improved likewise, and the time differences necessary for complete lateralization to one nostril became shorter.

Figure 89 gives some of the results, and shows that a time difference of 0.5 millisecond or less is sufficient to give a displacement of an odor sensation from a middle position to one nostril or the other. These results were obtained with several kinds of odor substances, whose effects do not differ much from one another.

Fig. 89. Localization and extent of smell sensation as a function of the time difference between two stimuli, one to each nostril. From *Journal of Applied Physiology* [112].

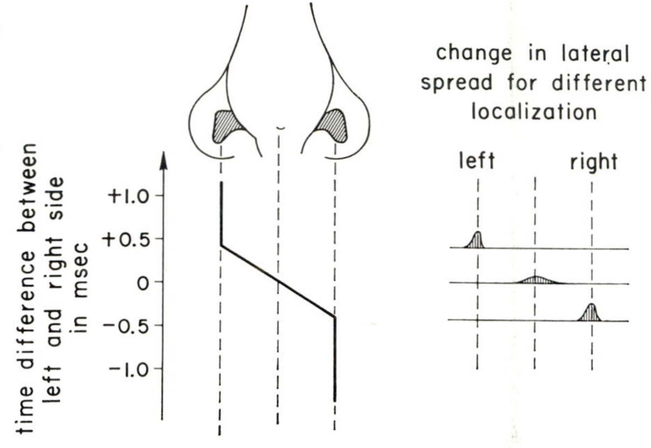

In other respects the odor phenomena are closely similar to the localization effects found in hearing. Figure 90 gives the results obtained with simultaneous stimulus presentation and a variation of odor concentration in the stimulation of one nostril.

Skepticism concerning the short time delays obtained in these experiments led to further tests. A hollow sphere about 1.5 inches in diameter, with an opening of 3 mm at one place was loaded with a dilute odor by first forcing pure humidified air through it and then introducing for 1 second a mixture of air and odor. This ball was presented to the observer in a free field in which stray odors had been eliminated, and it was found that a displacement of the sphere out of the midline could be appreciated if this displacement amounted to 7-10° at a distance of 8 cm from the nose (Fig. 91). It is evident that the time differences involved in the conveyance of the odor to the nostril from two angular positions at this distance must be extremely small.

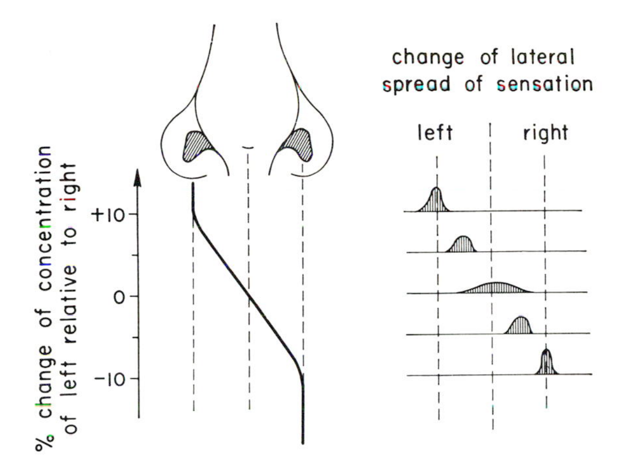

Fig. 90. Localization and extent of smell sensation as a function of an imbalance in the intensity of the stimuli in the two nostrils. From *Journal of Applied Physiology* [112].

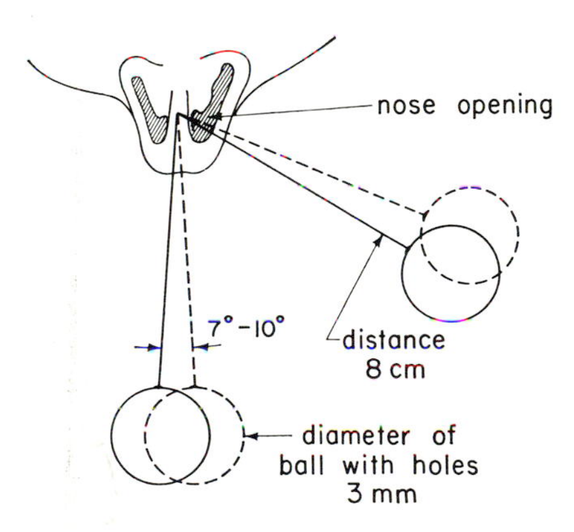

Fig. 91. Accuracy of localization of an odorous object with the sense of smell. From *Journal of Applied Physiology* [112].

Other tests were made in which the gaseous odor in the spheres was replaced by smoke and the movements of the smoke particles during inhalation were observed under strong illumination.

It is of interest to add that the pig, an animal well known to be talented in the discovery of truffles a little below the surface of the ground, has nostril openings that are widely separated;

when searching for such delicacies it constantly moves the head from side to side in the same manner that people do when trying to locate a sound source. It is likely that the animal is making use of its capabilities of odor localization.

The two components of neural activity

In contrast to the evidence for rapid funneling in the process of localization we are faced with the fact that sensory magnitudes such as the loudness of a sound require a good deal of time to develop their full value. As may be seen in Fig. 92, a period of about 0.2 seconds is needed for a tone to reach its maximum loudness, after which the loudness begins to decline as a result of adaptation. These measurements were made by presenting a standard tone pulse of 0.2 seconds duration and matching its loudness with that of other tone pulses of shorter or longer durations by changing their sound pressures.

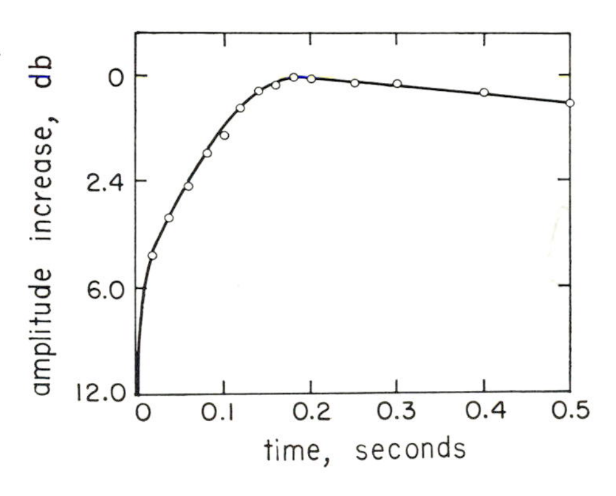

Fig. 92. Increase in loudness of a tone as a function of stimulus duration. From *Physikalische Zeitschrift* [3].

The ordinate of the figure represents the necessary increase in pressure amplitude in decibels for the different durations to make the loudness match. For vibratory sensations the rate

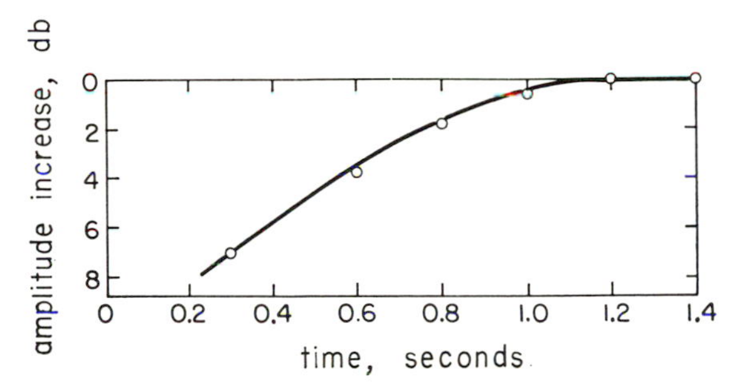

Fig. 93. Increase in the magnitude of vibratory sensation on the skin as a function of stimulus duration. From *Annals of Otology, Rhinology, and Laryngology* [118].

of development of sensory magnitude is much slower, and about 1.2 seconds is needed for a maximum (Fig. 93). For taste sensations, observations were made by Bujas and Ostojkit in 1939, and again about 1.2 seconds was required for maximum magnitude (Fig. 94). For smell, observations were made by equating to a standard a series of odor stimuli varied in

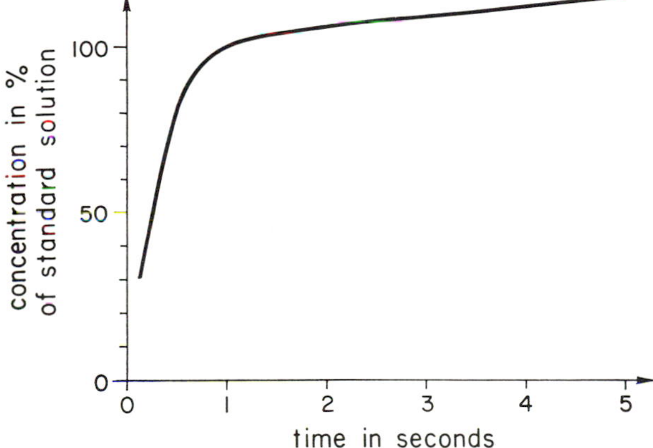

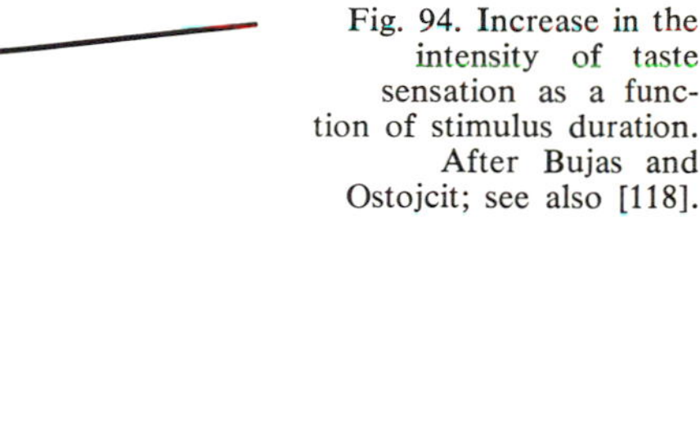

Fig. 94. Increase in the intensity of taste sensation as a function of stimulus duration. After Bujas and Ostojcit; see also [118].

duration and concentration, with results given in Fig. 95. Thus it appears that the development of sensory magnitude is a slow process as compared with the process of localization. There seem to be two processes in the nervous system,

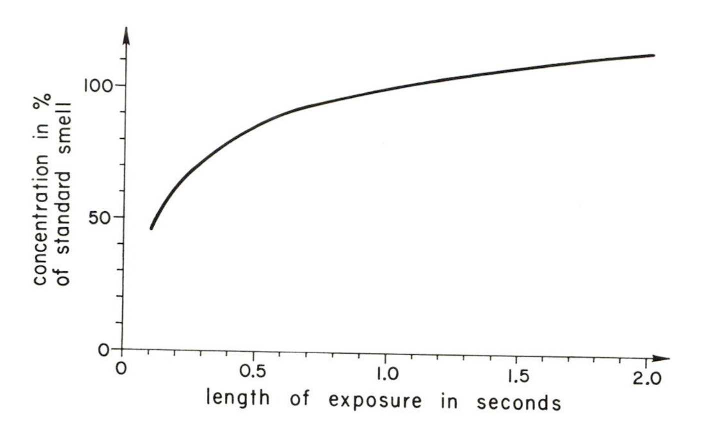

Fig. 95. Increase in the intensity of odor sensation as a function of stimulus duration. From *Journal of Applied Physiology* [112].

one concerned in localization that is completed in a few milliseconds, and another producing the sensory magnitudes and requiring more time, amounting to about 200 milliseconds in hearing and more than 1000 milliseconds in taste, smell, and vibratory sensibility. The time relation between these two neural activities is of the order of 1 to 200 or more.

This relation can be demonstrated more directly in an experiment in which two stimuli that are exactly equal in onset, duration, and cessation are localized. Under these conditions the localization is quite sharp, as may be seen in the upper portion of Fig. 96. If the duration of one of the two stimuli is reduced to half that of the other (lower portion of Fig. 96), there would be no change in localization if the onset time is the determining condition. But if sensory magnitude plays a role, and the center of magnitude of the pattern is determinate, then it would be necessary to delay the second stimulus (B') by some tenths of seconds to make it

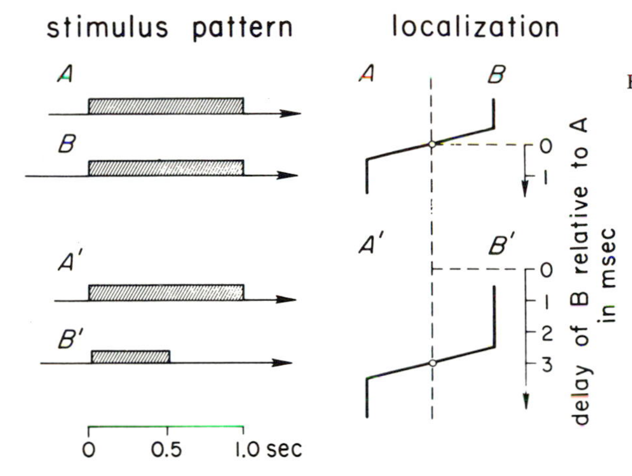

Fig. 96. The dependence of localization on the onset of stimulation. From *Annals of Otology, Rhinology, and Laryngology* [118].

equal to the other. The experiment showed that the necessary delay was only about 3 milliseconds. Hence there is only a little effect of sensory magnitude. This result shows that localization is determined in a short time—a few milliseconds—after the onset of a stimulus.

It is significant to note that the earliest recording of action potentials in a single gustatory nerve fiber, done by Pfaffmann (1941) in the chorda tympani nerve, showed at the onset of stimulation a sudden rapid burst of nerve discharges after which the discharge rate settled down to a lower rate (Fig. 97). Nearly all the records of sensory nerve discharges exhibit this initial burst. My conclusion is that the bursts at onset determine the localization of the sensation and later discharges determine such qualities as sensory magnitude and timbre.

Fig. 97. Single unit discharges for the nerve of taste. After Pfaffmann (*Journal of Cellular and Comparative Physiology*, 1941, 17, 243-258).

It is likely that the rapid action involved in localization follows a particular neural pathway, which then becomes refractory and is not accessible to the neural activity that follows. The later activity must be carried along other fibers, and these others represent the magnitude. The conclusion is that there are two steps in sensory nerve activity, one taking place in a few milliseconds and another requiring times of the order of a second.

One way of testing this two-step hypothesis is to present stimuli that lack a sharp onset. If localization is connected with the onset action then it should be significantly impaired if the stimulus is faded in slowly. It is well known in audition that it is easy to localize sharp clicks, which have a small acoustic image and well-defined contours. If the click is modified by passing either the electrical pulse or the acoustic wave train through a suitable filter, the sound image is made to seem larger and more diffuse. Then the location of the image becomes difficult to recognize. If the stimulus contains only low frequencies and the onset takes 10 milliseconds or longer, the sound image can spread over an angle of 20-30° and the localization is very poor.

The same experiment was carried out with a 5000-cps tone. For such high tones localization by phase changes no longer operates, because the nervous system is not able to follow the temporal pattern. However, such tones can be modulated or tone bursts can be used with various onset rates. When such tone bursts have sharp onsets the localization is easy, when they have onset times of a second or so there is no localization at all and the sound image seems to occupy the whole space in front of the observer.

The same results were observed with other sense organs, so that in general it appears that localization is determined at the beginning of neural activity and is independent of later occurring nerve discharges that determine other aspects of the sensation.

For two sensory qualities, warm and pain, I was unable to obtain a lateralization with time differences as short as 1 millisecond. These required differences of the order of 100 milliseconds (Békésy, 1962). The reason in my opinion is that with the techniques used the time of action of the stimulus was slow. The transmission of heat through the skin is so slow that the sense organs are changed in temperature very slowly. The situation therefore resembles that of a tone with a long time of onset, for which large time differences are necessary for definite localization. With an improved technique I hope to reduce the stimulus onset time. For pain sensations also the onset time is long, and too, it is difficult with pain to control the stimuli so as to produce two sensations that are equal in magnitude, which is one of the requisites for the investigation of temporal relations.

Factors of uncertainty in research

In experimentation and in theoretical developments also it is often surprising how difficult it is to obtain results of the necessary exactness. It is perhaps even more disturbing to discover how difficult it is to reproduce results that are wrong. To be sure, no observations are completely correct or completely in error, as is shown by many developments in the history of science. This situation emphasizes the importance of a consideration of uncertainty in research. Every type of research has its own level of uncertainty, and it is essential to ascertain the limits imposed upon the results.

In psychological determinations one of the sources of uncertainty is the observer. Only within limits can we accept an observer's statements as factual, or free of bias, or repeatable. Hence we must distrust the observer to a certain extent. One way of avoiding the difficulties here is to use several observers. If they fail to agree we can use statistical methods in handling the results. Unfortunately, however, we ought to distrust statistical methods also.

One of my worst experiences was when I had a large amount of data and asked a statistician how to treat it, how to evaluate it, and how to arrive at conclusions from it. To my surprise his first question was, "What do you wish to prove?" The fact was that I did not wish to prove anything, but only to find out what the data showed and what new relations could be formulated from them. Certainly there are instances in which it is easy to do this, but on the other hand there are many situations in which statistical manipulations are of no assistance.

In general it is believed that electrophysiological observations are free of error and more objective than psychological observations, but I am not sure that this opinion is justified.

In work on a problem in electrophysiology a great many electrical responses might be recorded and might occupy a few hundred feet of photographic film. The experimenter must then decide which foot of the total is the part to publish. The selection of a small section out of the total record involves a psychological judgment of great importance.

There are a few situations in which, if an experimenter asks someone to select more or less at random the particular section of the extensive record to focus attention on, the results thus noted can be reproduced by any other section. For the most part this situation occurs only when the conditions are relatively fixed and the organisms are simple ones like insects. For the more complex living systems this degree of uniformity is seldom found.

Another difficulty in the field of electrophysiology is that nearly all recordings are a function of time. It can happen that for a particular neural activity the factor of time is much less important than the one of space. The techniques now available are not well suited to the simultaneous recording of actions at a large number of electrode positions, but I am sure that in time a significant advance will be made in this procedure.

A third difficulty in electrophysiology, and perhaps the most serious one, is that we obtain at the end of an experiment a record that must be evaluated by eye, and this kind of evaluation may be vastly different from the evaluations that nervous systems are capable of making. For example, some years ago I recorded some samples of speech on magnetic tape, and on the same tape made a clear oscillographic record of the sound pressures. With training it was possible from viewing the record on an oscilloscope to guess what the speaker was saying. I met someone who was able to read the Turkoman languages and who had extensive training in Chinese and Japanese writing, and I asked him to read speech in these languages visually from the oscillographic display. It turned out that he did very well in deciphering what the speaker was saying.

For these recordings the microphone was a foot away from the speaker. In further experiments I increased the distance to 20 ft. I then asked the Chinese expert whether this change affected his understanding of the speech when he listened to it, and he was certain that it made no difference. Yet when the oscillographic display was presented to him he was totally unable to decipher it. He could only make out a few vowels like "a," "e," and "u." The difference thus demonstrated between a visual and an auditory perception of the same speech record arises from the different temporal patterns picked up by the two nervous systems.

Another difficulty or uncertainty is in the selection of the type of animal to use in experiments. Simple animals may give clear-cut results, and in their use there is no trouble in getting repeatable physiological records. The question is whether an animal such as a blowfly or a *Limulus* does the same things that we do.

A new approach to the problem of the operation of the nervous system is to make a model and investigate it. There are many ways of making a neural model, and with practice we can produce a sort of non-Euclidean nervous system that is correct within itself. But relating the model to the actual nervous system introduces uncertainty. At present this new method is valuable because it suggests what facts and relations ought to be observed and investigated. It might happen that the results produced by the non-Euclidean nervous system would be exactly the same as those produced in the nervous system of man, and yet the method of producing these results might be different, so that we would still not have learned anything about the human nervous system even though the model accomplished the same things.

As there are many possibilities of constructing a neural model there is little chance at present that any model acts internally in the same way that man's nervous system does.

Apart from this uncertainty there is always the uncertainty that is produced by the physical facts, which is known as the Heisenberg principle.

With all these sources of uncertainty in mind, it becomes clear that the best method of reducing or eliminating them is to approach a given problem by different methods. Then we can compare the different solutions, and in noting the common features in the results we increase to some extent the probability that the answer is a correct one. This is why, in the lectures to follow, I wish to emphasize the common features found in results obtained especially in the fields of inhibition and nerve transmission.

Central inhibition and the observer

Because electrophysiology is not perfectly safe as a source of understanding of sensory phenomena, we need psychological observations as well. A demonstrated agreement between electrophysiological and psychological data ought to improve greatly the reliability of the results.

Here we come to the questions of how to make good psychological observations and how to make electrophysiological observations that are not affected by psychological factors.

To make good psychological observations certain conditions are required: (1) We need well-defined stimuli. (2) We need a method of stimulation that yields sensations that remain relatively constant when the psychological conditions are the same. (3) We need to have a clear conception of the different kinds of sensation that may arise from a particular stimulus. (4) We need observers who can inhibit sensations that are irrelevant to the situation.

To show how difficult it is to produce a well-defined stimulus I shall discuss the matter of the mechanical stimulation of the skin surface. Initially the problem seems simple. No surgical procedures are required; there are no special precautions. We simply place a stimulating device in contact with the skin. The sense organs within the skin are clearly able to respond to a deformation. Deformation and pressure are related by the impedance of the skin surface, which is a simple physical quantity whose magnitude is easily determined. In the experiment illustrated in Fig. 98, the classical observations of Meissner (1859) were repeated. Meissner found that when a finger is immersed in a vessel filled with mercury no pressure is felt at the fingertip but only at the mercury surface where there is a change in the steady pressure applied to the skin. This evidence shows that no sensation arises in the finger from pressure as such, but only when there is a pressure change, a displacement produced by the pressure gradient.

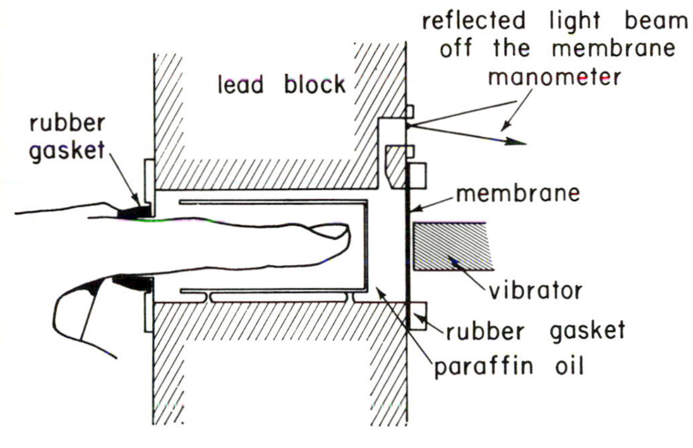

Fig. 98. A repetition of Meissner's experiment with vibratory pressures. From *Akustische Zeitschrift* [33].

Meissner's experiment was repeated by the use of alternating (instead of steady) pressures by the method shown in Fig. 98, and the results were exactly the same. It is a pressure difference between two points on the skin that stimulates the end organs.

How difficult it is to determine what the physical stimulus is even in the simplest situation can be seen from Fig. 99. Here is a schematic

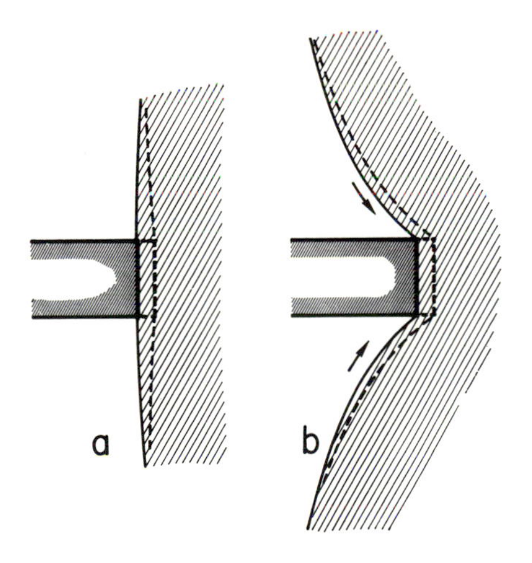

Fig. 99. The effect of a static deformation of the skin on the generation of shearing forces by a vibratory stimulus. Two different levels of pressure were used, as shown at *a* and *b*. Shearing forces become significant when the level is high. From *Akustische Zeitschrift* [33].

representation of the effect of pressing a dull needle against the skin surface. The resulting deformation can consist not only of a displacement parallel to the axis of the needle but also of a shearing displacement in a radial direction. The shearing displacement produces a pressure gradient of a complex nature perpendicular to the skin surface.

Great progress in stimulating the skin surface was made by von Frey (1914, 1926, 1929) in his introduction of the test hair known by his name, already shown in Fig. 25. With this instrument we can produce a well-defined pressure at a point on the skin, and we can keep this pressure constant. We do not know about the shearing forces because they depend upon the properties of the skin.

The action of a von Frey test hair is pictured in Fig. 100. It provides a unique direct pressure stimulus because when the hair is pushed downward to a certain extent its pressure reaches a maximum and remains constant thereafter, as the figure shows. The distance between the end of the handle in which the hair is fixed and the surface of the skin can be varied considerably without altering the pressure; the instrument can be held in the hand, and the involuntary movements of the hand will not affect the direct pressure. The principle of the method is that a beam that is held vertically and pressed against a rather stiff surface will increase its pressure until the beam bends, as shown, and thereafter as the holder is moved farther the pressure will not increase, provided that the tip of the hair is free to rotate about its center. It is essential that the beam be applied perpendicularly to the surface, for otherwise the pressure will increase

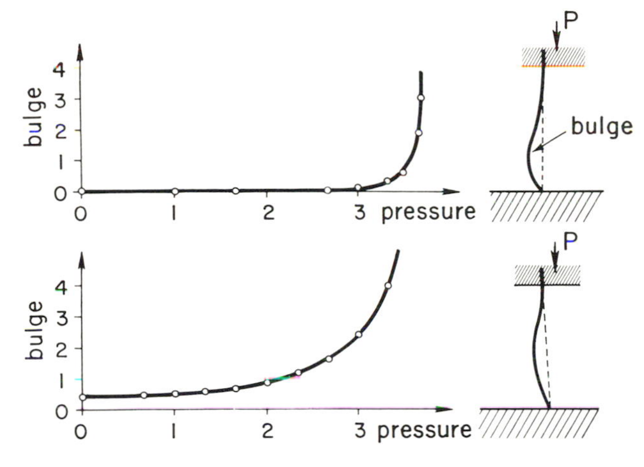

Fig. 100. Pressures produced by a von Frey test hair. From *Journal of Experimental Psychology* [37].

continually (lower part of Fig. 100). By changing the length, radius, or elasticity of the beam we can alter the maximum pressure obtained after the beam bends. The theory of such a beam has been developed in the study of the supporting strength of beams.

As shown earlier in Fig. 13, a vibrator applied to the skin produces traveling waves along its surface. These waves are of great complexity because the skin is not a homogeneous structure but a rather stiff membrane attached to soft layers. This type of wave is greatly dependent upon the vibratory frequency. In physical terms, such waves exhibit dispersion. Hence an elaborate formulation is required to specify the speed of the surface waves, their damping, and their penetration into the deeper layers of the skin. If the skin surface shows changes in a lateral direction, this change may produce complex reflections. I have always hoped that geologists in the study of earthquake waves would throw light on the nature of Rayleigh waves (see Rayleigh, 1945) and Lamb waves (see Lamb, 1932), but now it appears that the geologists might learn something by observing waves on the skin.

All these surface waves transmit periodic forces into the deeper layers of tissue, and can even reach the bones of the arms and legs. The transmission of vibrations by the bones occurs at a high speed, and produces added complications when the activities of the bones are conveyed back to the skin. There are some very sensitive end organs near the joints, so that the experience is complicated by the appearance of vibratory sensations there. The joint sensations may interact through summation or inhibition with the effects produced in the skin in the immediate region of the vibrator.

Figure 101 shows how the bones of the arm can be set into vibration by a stimulus applied to the hand. Because the bone vibrations are but slightly damped (see Fig. 102), they may produce a considerable alteration in the vibratory threshold.

A persistent question has been whether vibrations are sensed by endings in the skin or in the joints. This question can be approached experimentally by conveying vibrations to the spine. This may be done in two ways. One way is by lateral displacements of the hip as in Fig. 103. At certain (resonance) frequencies the whole spine can be made to vibrate and the observer can be asked to state where the vibrations are felt. The experiment is a difficult one. A second method is to apply an electrodynamic vibrator to the spine near its end where the skin is in contact with the bones. Then it is possible to

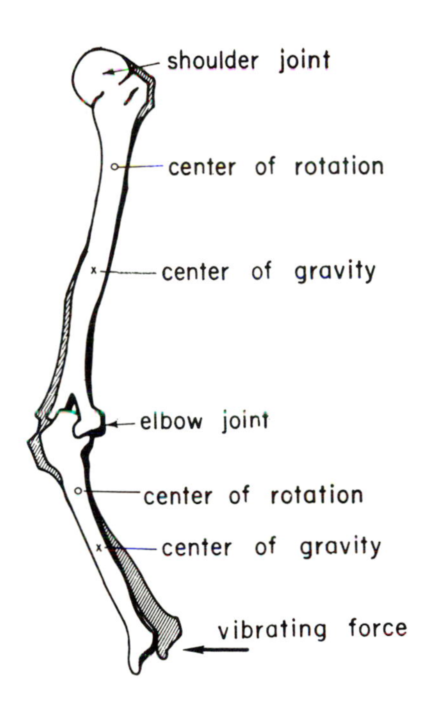

Fig. 101. Transmission of vibrations through the bones of the arm. From *Akustische Zeitschrift* [33].

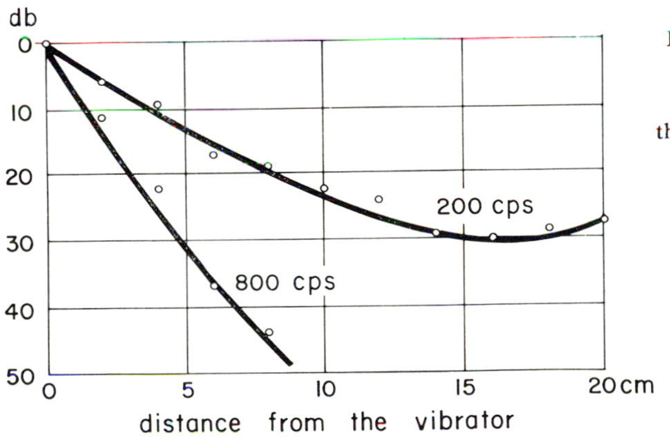

Fig. 102. The ordinate shows, for two different frequencies, the decrease in decibels of the amplitude of vibrations as measured on the skin of the arm. From *Akustische Zeitschrift* [33].

Fig. 103. The resonance characteristics of the spinal column as a function of the frequency of lateral vibration. Redrawn from *Akustische Zeitschrift* [34].

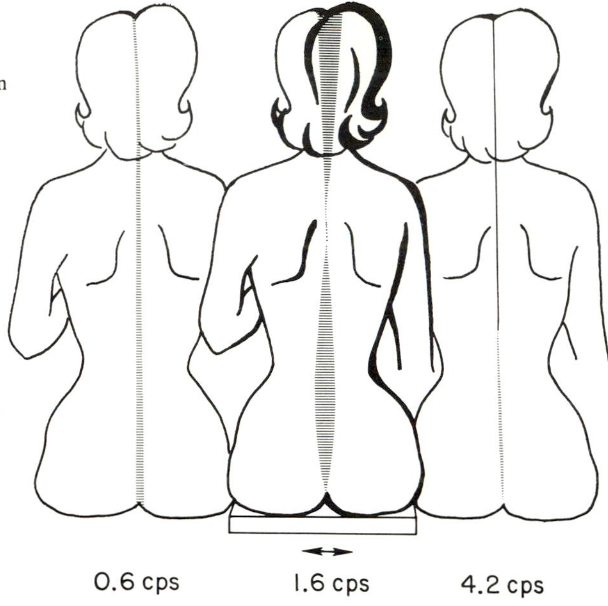

cause the spine to vibrate without producing much vibration on the surface of the skin along the trunk. It is a peculiar experience to have an impression of the spine in vibration, in isolation from the remainder of the body. This was the first time that I actually felt my whole spine by itself. The damping along the spine is only moderate, and for low frequencies the vibrations can be perceived from hip to head.

The above procedures demonstrate the difficulties involved in stimulating the skin by itself, and show the common presence of wild types of traveling waves. The situation is similar to that encountered in stimulating the skin with electrical waves by the use of a small point electrode and a large indifferent electrode else-

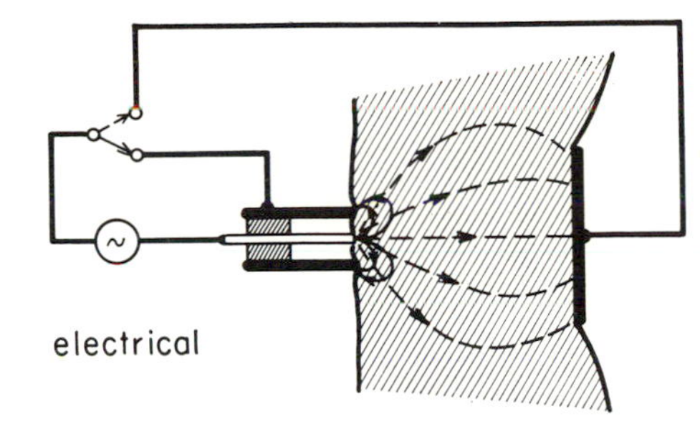

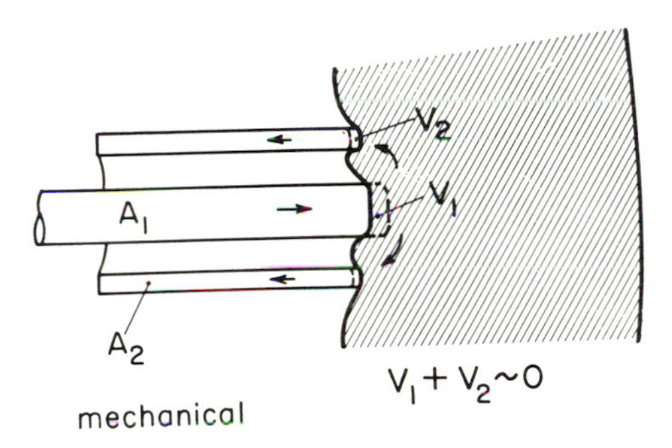

Fig. 104. Method of minimizing the spread of electric currents and mechanical vibrations by the use of a counteractive ring. From *Journal of the Acoustical Society of America* [85].

where, which is the usual method of electrical stimulation. A better form of electrical stimulation is obtained by the use of a concentric electrode, as shown in Fig. 104, which makes it possible to reduce stray currents and to define the locus of the stimulus.

The satisfactory results obtained with concentric electrodes led to the development of a mechanical vibrator with similarly defined characteristics. The construction is shown in Fig. 105; the tip of the vibrator (V_1) is surrounded by a tube that is also vibrating but in an opposite phase in relation to the tip. The amplitudes of both vibrators are adjusted so that the volume displacement on the skin surface is the

Fig. 105. Method of producing the mechanical analog of a concentric electrode. From *Journal of the Acoustical Society of America* [85].

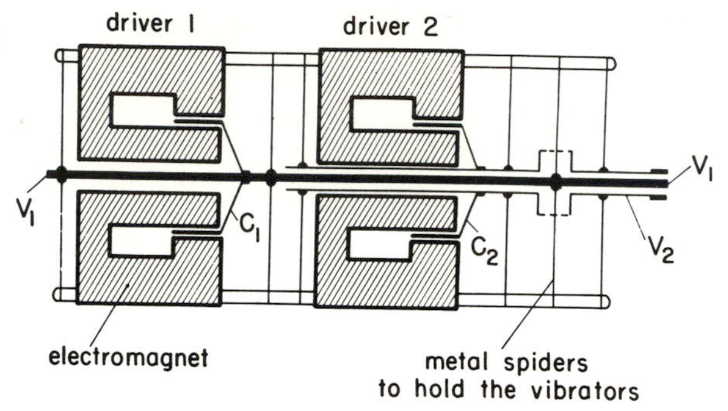

same for each. Therefore, when the skin is touched by both vibrators at the same time the total volume displacement of the skin becomes zero, and there is no pressure acting on other (lower) levels. The principal mechanical stimulation occurs in the skin between the two concentric ends.

Unfortunately, as Fig. 105 shows, two vibrators with independent adjustments of amplitude and phase are required. The setup is very complicated, heavy, and awkward to place in proper position. But for vibratory research on the skin surface it is the type of apparatus that ought to be used. This discussion of the mechanics of stimulation may not seem pertinent to the subject of inhibition, but I feel it necessary to point out that many of the criticisms that have been made of observers should rightly be directed to the poor understanding by the experimenter of the physical aspects of the stimulus. I have found little in the handbooks of physics that would assist in the specification of the stimuli that are dealt with in psychology and physiology. The reason is that only homogeneous media can be treated mathematically without becoming irrational.

The problem of precision

It seems unhappily true in sensory psychology that either it is difficult to define the stimulus properly in physical terms or the sensations that are to be observed are complex even when the stimulus is simple. In hearing, for example, the sound pressure at the entrance of the external auditory meatus is easily measurable and well defined. Some difficulties appear if we wish to define the stimulus at the eardrum, and they increase further when we try to specify the stimulus at the level of the hair cells of the cochlea. We still can neglect the details in psychological measurements and use the pressure at the meatus as a reference.

The sensory effects produced by a pure tone are so numerous that observations are difficult to describe. A pure tone has pitch, loudness, tonal volume, tonal density, direction, distance from the center of the head, smoothness, a complex group of overtones (timbre), and changes in all these attributes during the time of presentation of a stimulus. An observer must be well trained to be able to concentrate on only one of these attributes, and such training takes a long time.

In spite of these complications, some clear-cut relations between physical stimuli and sensations have been found. The first law of psychophysics was discovered by the Pythagorean School in 300 B.C. This law stated that the length of a plucked string is related to the pitch: reducing this length to half will raise the pitch one octave. The law included the statement that such a change in the length of a string will produce this result regardless of the initial length or of the tension in the string. The law was so generally accepted that now it is no longer called a law. A reason for this is that probably side effects of the stimulus, such as changes in the overtones and in loudness, do not seem to affect the judgment of pitch.

Another problem is that of maintaining constancy in the phenomenon under study. It is necessary to investigate fully all the variables that can affect the phenomenon. This investigation can take more time than the measurements on the phenomenon itself. I am always surprised at the mistakes that are made in estimating the importance of certain of the variables.

As an illustration, a record was made of changes in pulse rate during the time that an observer was making some matches of pitch. In a period of 4 minutes the rate of the heartbeat declined to half its former value (see Fig. 106, which represents a cumulative record). Yet over the same period the matching of pitch remained constant with a precision of 0.1 per cent. It appears that the blood flow has little effect on pitch adjustments. However, other phenomena are strongly affected by an oxygen lack. Directional hearing, for example, is greatly altered when an observer breathes a mixture of nitrogen and air that contains less than the normal amount of oxygen.

Another methodological consideration is the number of types of sensation that a stimulus can produce and how different they are. Everyone is familiar with the drawing of Fig. 107 that can be interpreted in several ways, for example, as a flat frame, a truncated pyramid approaching the observer, or a truncated pyramid

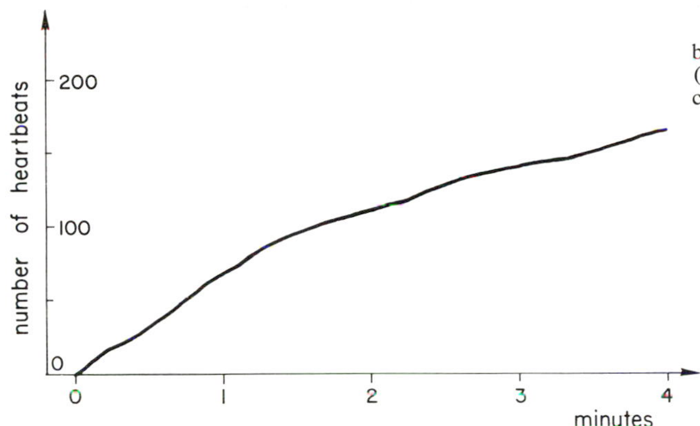

Fig. 106. Cumulative record of the heartbeats of a human subject. (For a constant rate the curve would be straight.) From *Journal of the Acoustical Society of America* [110].

facing away from the observer. Since there are no secondary cues in this drawing to indicate which of these possibilities is correct, we have a multiple-choice situation and observers will differ widely in their observations unless some direction is given to them for their guidance. If the observers disagree it is no fault of theirs, but of the programmer of the experiment. The best way to obtain agreement in such a situation is first to allow the observers to discover the three possibilities independently and then to compare their observations. In some situations this is the only means of arriving at agreement.

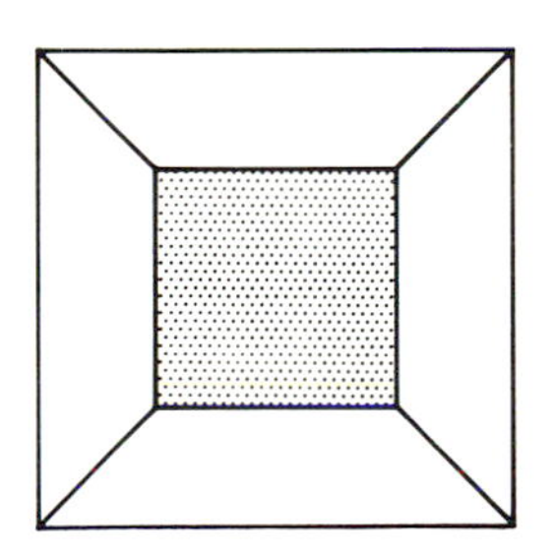

Fig. 107. Ambiguous figure.

For the fun of it I should like to tell a short story. At the time when I was a beginning communications engineer earphones were always used in listening to the radio. One day an important politician came to the laboratory bitterly complaining that while listening to music on the radio he heard the entire orchestra in back of his head. His wife had a solution to the problem, which was to visit a psychiatrist. This would have had serious political consequences, and instead he asked me to repair his radio. The problem, which was solved in a two-hour session, is illustrated in Fig. 108. If a person has an earphone on each ear and the two are well matched the acoustical situation provides a free choice of localizing the imaged sound source in front, within the head, or behind. The determining condition will probably be some early experience relative to this situation.

Fig. 108. The ambiguous localization of a sound listened to with a pair of earphones.

For the patient's cure he was asked to sit in front of a sound source, which he localized correctly in front. Then he was asked to put his two hands in front of the ear openings and to flap them back and forth. After a while, with the proper hand positions, he was able to hear the sound in back of his head, and then he could make the sound change its position from one place to the other. We then used the visual situation of Fig. 107 to explain that he had three choices in his perception of the figure. In a short time the person had no difficulty in locating the orchestra in front or behind, as he pleased, and his problem was solved.

Under some conditions a multiple-choice problem can be so complicated that it is difficult to obtain agreement even among well-trained observers. As an illustration consider the localization of four clicks, two in each ear, with time relations as represented in Fig. 109. There are four ways in which the click pattern can be perceived. One way is that the first two clicks presented to each ear are the only ones perceived and the delayed ones are completely inhibited. Another way is that the first clicks are inhibited and only the last two are used to locate the sound source. There are two further possibilities involving a combination of the first click in one ear with the second click in the other ear. When we begin to change the relative delay times of these clicks the localization can vary in four different ways as shown in Fig. 110.

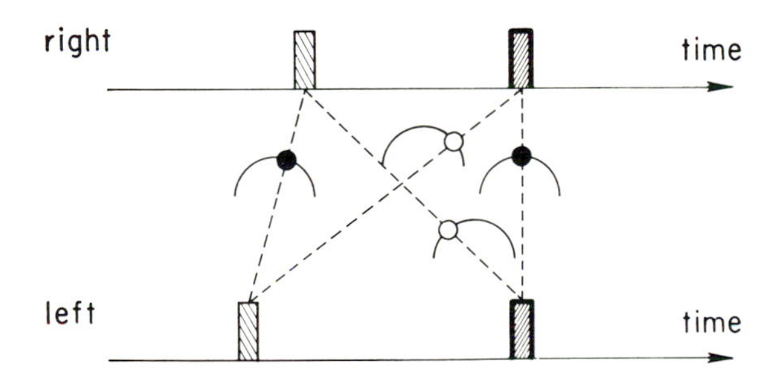

Fig. 109. The types of sound image produced by two pairs of clicks.

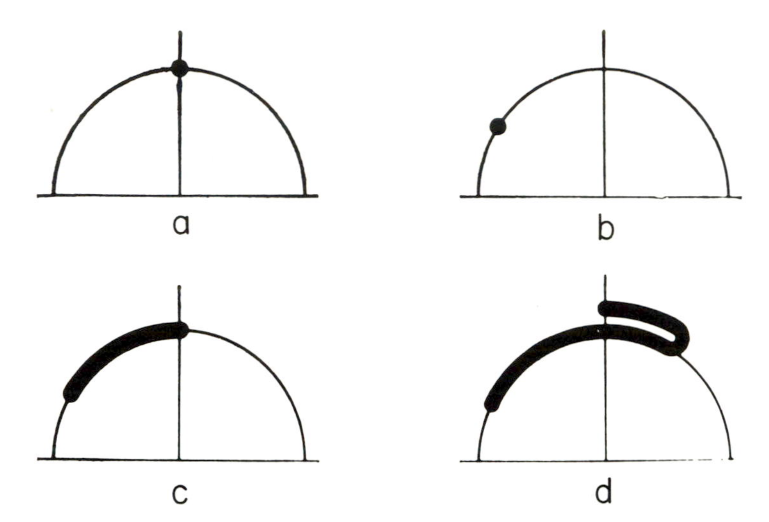

Fig. 110. Paths of movement around the head of the sound image produced by two pairs of clicks, under four different conditions. From *Physikalische Zeitschrift* [5].

In my experiment on this problem three or four months were required before the whole situation was mapped out and the five observers could agree with one another. The best way of assisting the observers in discovering all the possibilities was to use filters to change certain pairs of clicks a little so as to emphasize them.

The next problem is the selection of a good observer. Some so-called good observers report what they think is expected of them. I call this one the destructive type. They are present in large quantity in any laboratory that has a theory in one field or another. In my opinion a good observer is one who after long training can inhibit some of the phenomena that are not directly under observation.

I worked for a long time on the problem of the best acoustics for studios and concert halls. The crucial question was who should decide when a concert hall was at its best. As a physicist I could produce any kind of reverberation time within a reasonable range. I could place emphasis on certain ranges of frequency, and I could alter the ratio between the direct sound coming from a musical instrument and the reverberation from the walls of the room. With these measures I had done my duty as a physicist and it was the responsibility of the musician to indicate what was wanted. We had about ten experts who made the judgments independently. I had the impression that some of the most famous musicians were unable to make the kind of observations that psychological observers are accustomed to make. Yet it was difficult to make the point officially that certain famous persons should not be involved in the planning. After a struggle the following method was adopted.

We asked each musician to play his favorite instrument so that it sounded best to him under each of various conditions that were produced by changing the sound absorption on the walls of the concert hall. We placed a small condenser microphone or a vibratory pickup inside his instrument and recorded the musical passage. This passage was an easy melody, which was repeated for each performance. Figure 111 represents two records of the amplitudes obtained by a noted conductor under two conditions for the playing of Chopin's *One Minute Waltz* on the piano. It will be seen that when the room was reverberant the piano output was reduced. Then in a highly damped room the pianist put forth all his power to bring the sound to the level to which he was accustomed. This experiment made it clear that he was well aware of the acoustic effects of the reverberation.

Fig. 111. The acoustic level of performance of a piano player under two conditions of room reverberation. From *Forschungen und Fortschritte* [30].

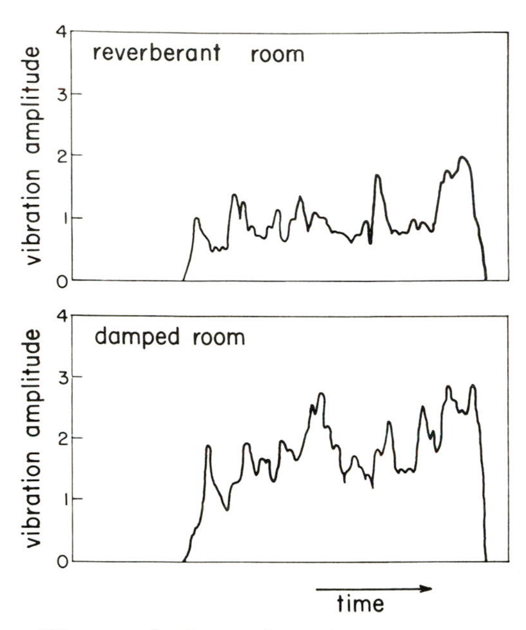

We tested the technical ability of this performer by going to a very difficult piece of music. Again the records showed the same adjustments to the changes in reverberant conditions. He was not disturbed by the technical difficulties of playing while observing the reverberation effects, and we considered him well suited as an expert on room acoustics.

The same experiment was repeated with other expert musicians. Many did not perform as well as the one just mentioned. About half of them played their instrument with the same intensity under all conditions of reverberation. Clearly they were not observing the phenomenon for which they were asked to be experts. Others did well in adjusting the performance of simple scores, but failed in playing at sight a score with which they were not familiar. They were good observers but were limited in technical ability in their playing.

It may seem surprising that neural processes should show the precision that they do in certain phenomena of localization, in which a time delay of as little as 1 millisecond can be detected by a displacement of the sound image from the median plane of the head. Yet other functions of the human system exhibit reactions of the same order of precision. For example, an expert runner will lift his foot about 40 cm, but he must put it down with a precision of about 4 mm to avoid striking his toes in a manner that might break them. The coordination of the leg movements under these conditions requires a precision of about 1 per cent and involves not only sensory perception but also an interaction between perception and muscle performance in a very brief period of time.

It is obvious that the simpler the system is the more precise will be the sensations and reactions. We expect the precision to be different for the different sense organs. The production of nerve spikes in a single fiber with its all-or-none characteristic should be relatively stable. The sense organs that operate according to the compensation principle as discussed earlier are probably more sensitive and necessarily less precise. When inhibition plays a major role, the precision will not be expected to be great. Nevertheless, it should be remembered that the sensation of the pitch of a 1000-cps tone can be adjusted to a precision of 0.3 per cent, even though inhibition is probably involved.

From this discussion I should like to conclude that a good observer is one who has the ability to inhibit the unnecessary phenomena. Yet it should be mentioned also that in the design of experiments it is important to be sure that the observations expected of an observer are really observable. To illustrate this point in an exaggerated way, let me mention the fact that it is difficult to believe that the only way one can observe the widening of the pupil of one's own eye is by looking in the mirror. Evidently the eye has no interest in evaluating the light intensity on the basis of the size of the pupil.

IV · STIMULUS LOCALIZATION AS A METHOD OF INVESTIGATING NEURAL ACTIVITY

Directional hearing has always seemed to me to be a useful tool for research in physiology and psychology. The precision is great if a stimulus is used whose intensity has a sharp onset. This requirement can usually be met, and then localization becomes a sort of stopwatch that can be used to determine short time delays in physical or physiological systems. Time delays less than a tenth of a millisecond can be detected with high accuracy if proper precautions are taken. It is essential that care be taken to make the magnitudes equal for the two stimuli whose time relations are to be measured.

I first recognized how useful directional hearing can be when I was in the woods in Austria. Some of the roads took a perfectly straight course through deep, dark woods. I could not imagine how such straight roads had been cut through the forest when the usual optical methods used by road surveyors would seem to be useless. Also it is known that some of these roads are very old and were probably built before the introduction of the theodolite. I was told that some of these roads were laid out by an acoustic method. A man stationed at the starting point noted the direction of the sound produced by someone at the other end blowing a horn. The first man then walked toward the sound source, marking all the trees on the way. It turned out that this method produced a straight line from starting to finishing points.

It is also well known how in World War I airplanes were localized by methods based on directional hearing.

The speed of traveling waves along the basilar membrane

The internal stopwatch that the localization method gives us can be used to estimate the speed of a traveling wave in the human ear and to check some of the data obtained by direct measurements on this speed in human cadaver specimens. Measurements of the speed of traveling waves along the basilar membrane provide a good method of proving that the elastic properties of this membrane are the same for a living person and for a fresh and still warm cadaver. A click such as indicated by *a* in Fig. 112 is presented to the ear in each instance. The click will set the middle ear in motion, and this motion will be transmitted to the basilar membrane, along which will appear a traveling wave that moves from the basal end toward the apex of the cochlea. The basilar membrane, which in man is about 35 mm long and 1 mm wide, is shown schematically in Fig. 113. The traveling wave during its travel stimulates practically all the end organs resting on the basilar membrane.

According to the localization theory of hearing, the high frequencies produce a maximum of vibration close to the stapes at the basal end of the cochlea, whereas the low frequencies produce maxima mainly at the opposite end. Therefore a strong low-frequency tone applied to the ear will produce an adaptation in which the

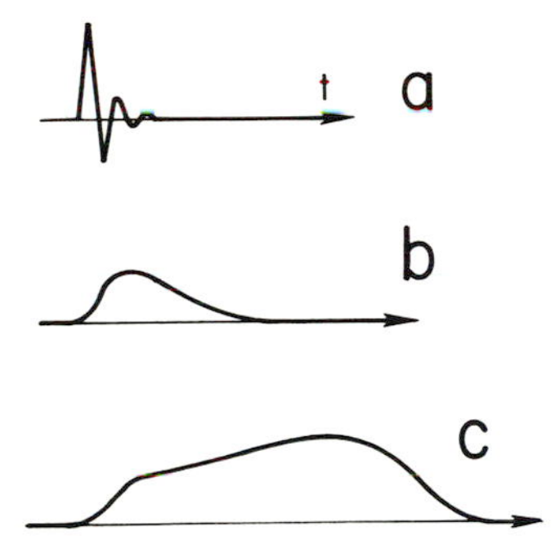

Fig. 112. In *a* is shown a click stimulus, and in *b* and *c* the time patterns of excitation after fatiguing with a low-frequency tone (for *b*) and with a high-frequency tone (for *c*). From *Physikalische Zeitschrift* [15].

sense organs at the apical end are made insensitive. Thereafter the traveling wave produced by a click will have an effect only in the initial region, with a pattern like that shown in *b* of Fig. 112. A strong high tone, on the other hand, will

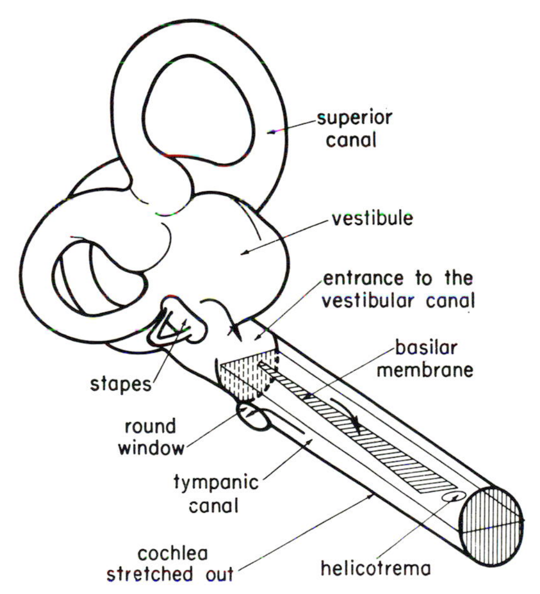

Fig. 113. Schematic drawing of the labyrinth with the cochlea uncoiled. From *Neural Mechanisms of the Auditory and Vestibular Systems* [92].

cause adaptation of a different kind, by reducing the sensitivity of the end organs in the basal region. Then the traveling wave caused by a click will have to travel a little distance (which will require a short time) before it reaches a section of normally sensitive end organs. The form of the pattern then produced is shown in *c* of Fig. 112. There will be a time delay in the excitation produced by form *c* as compared with that produced by form *b*. To observe the effect we need only to adapt the left ear with a strong high-frequency tone and to adapt the right ear with a strong low-frequency tone, and then present simultaneously to the two ears the same click stimulus. The click in the left ear then will seem dull, and in the right ear will seem sharp. The perceived loudness of the two clicks is then adjusted to equality by changing the amplitude in one ear, and the click stimulus is again presented to both ears at once. Then the sound will be localized in the right ear. In experiments to evaluate the effect precisely, the click stimulus was delayed in the right ear relative to its time of presentation in the left ear, and this delay was varied until the sound image seemed to be in the midline between the two ears. This procedure balances the effects of the different adaptations of the two ears by a change in the time of presentation in the two ears. The time delays necessary for this balancing are shown in Fig. 114 for tones from 50 to 1600 cps presented to the right ear and a tone always set at 1000 cps in the left ear. These times represent roughly the time of travel along the basilar membrane between the two adapted sections. If the frequency difference between the adapting tones is small the time delay is brief,

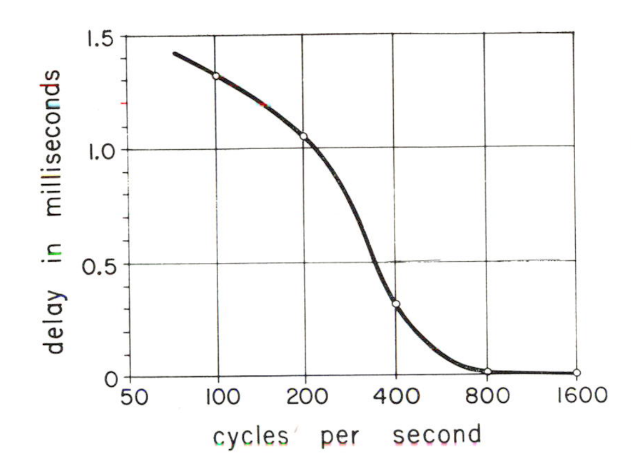

Fig. 114. Differences in the effective time of perception of a click after fatiguing the ear with various tones. The left ear is fatigued with 1000 cps throughout. Redrawn from *Physikalische Zeitschrift* [15].

and if the frequency difference is large the time delay is long.

Another method, not using adaptation, for showing that different tones involve different travel times along the basilar membrane will be described. Preliminary experiments showed that the presentation to one ear of a tone of 750 cps and to the other ear of a tone of 800 cps produced two separate sound images, one located in each ear. However, a complete (100 per cent) modulation of these tones at a rate between 5 and 50 cps, in which care is taken that the modulation occurs simultaneously and in phase at the two ears, will give a single sound image. The position of this combined image relative to the two ears is then of interest.

The experimental arrangement for further study of these relations is given in Fig. 115. The modulation rate was fixed at 8 per second, and could be adjusted for phase equality by means of a phase shifter. This adjustment was checked on an oscilloscope. The frequency of the oscillator in one channel was fixed at 800

Fig. 115. Circuit for study of the binaural interaction of modulated tones. From *Journal of the Acoustical Society of America* [106].

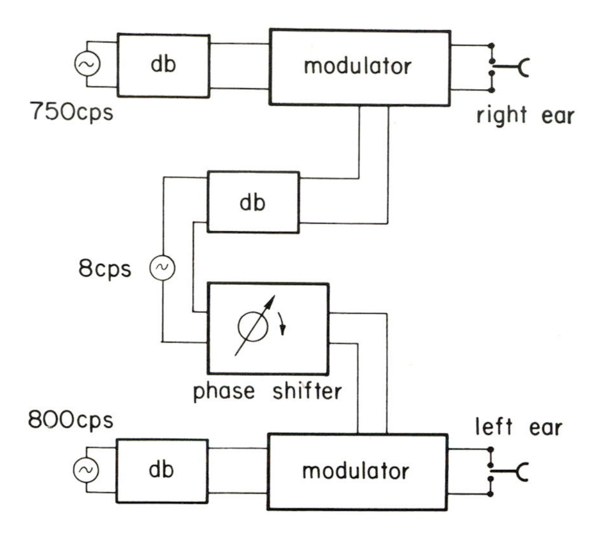

cps, and that in the other channel (shown as 750 cps in the figure) could be varied upward and downward.

When the frequency was set at 800 cps in both ears, the modulated image was located precisely in the midline between the two ears. When the frequency in the right ear was reduced, the sound image moved to the left side and it reached its extreme position of lateralization at 750 cps. This condition is represented in *a* of Fig. 116. On the other hand, when the frequency in the right ear was increased the sound image moved to the right and reached its extreme lateral position at 880 cps (*e* in Fig. 116). Intermediate frequencies gave smaller displacements as shown, but always the sound moved away from the middle toward the ear stimulated with the higher frequency. This fact indicates that the higher pitch is developed a little earlier than the lower one. This condition

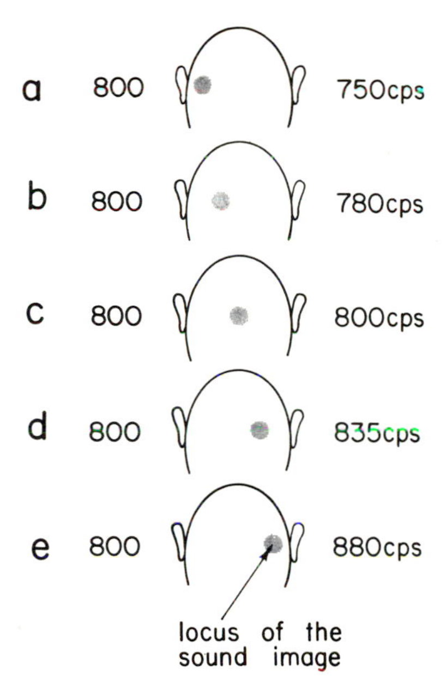

Fig. 116. Localization of modulated tones as a function of the frequency relation between the two ears. From *Journal of the Acoustical Society of America* [106].

is exactly what is expected according to the place theory of hearing, and indicates that the sensation of pitch is correlated with the distance of the place of stimulation along the basilar membrane as measured from the footplate of the stapes at the basal end, which is the starting point of the traveling waves.

The time delays can be measured by operating the phase shifter (Fig. 115) so that the sound image is brought back to the middle position between the two ears. Because lateralization in directional hearing is usually connected with inhibition, it is not surprising that during the displacements of the combined modulated sound image the pitch of the combination changes from high to low as the image in Fig. 116 moves from right to left.

The periodicity of neural activity after stimulation

Our increasing knowledge of the nervous system points more and more toward a complex network of afferent and efferent nerve fibers that seem to form certain loops, as shown in Polyak's drawing in Fig. 22. For an electrical engineer such a feedback circuit raises the question: Why does not this specific circuit begin to oscillate? Not long ago it was a difficult matter to construct an amplifier with high gain and a wide frequency band without unwanted oscillations. Such oscillations exist in biological systems, as for example the beating heart. Other oscillations arise under abnormal conditions, such as ringing in the ear or the oscillating scotoma described by Fröhlich (1921). Under normal conditions they are rarely observed. Oscillations in the visual system were described by Charpentier (1892) in the last century. Fröhlich (1913) made electrophysiological observations on the retina of the octopus, and Arvanitaki (1939) found oscillation in the giant axon of the squid. But in these experiments it is possible that the nervous systems were not in a completely normal condition during the observations.

To avoid difficulties that might arise from the question whether conditions were normal or abnormal the observations reported here were made psychophysically. An attempt was made to develop a method by which an oscillation or rhythm in the magnitude of sensation could be observed. The idea was developed in the following manner.

From the study of the localization of vibratory sensations on the skin we know that the localization of two clicks can be determined very precisely even when noise is present and even when a masking noise of different character is superimposed upon the clicks. It was a simple technical matter to measure coincidences with a precision of about 0.1 milliseconds by presenting one click to one ear and another click to the other ear. For the observation of physical coincidences under very unfavorable conditions I had the impression that I could do better with the ear than with the electronic equipment available at that time. In preliminary tests to evaluate this impression I presented to one ear a series of 50 pulses per second as shown on the left in Fig. 117 and a series of 10 1/3 pulses per second as shown on the right. Under these con-

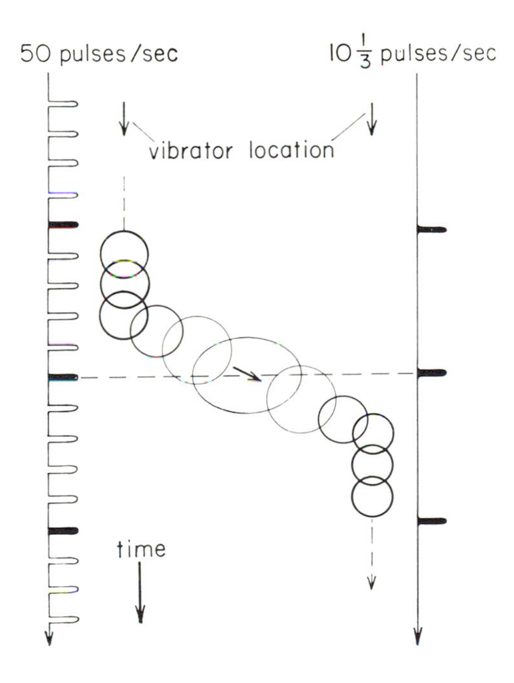

Fig. 117. The effect of extraneous stimulation on the localization of vibratory stimuli. From *Journal of the Acoustical Society of America* [83].

ditions the left ear had about 5 times as many pulses as the right ear. On listening to these two click series, or by using them to actuate two vibrators applied to the fingertips, it was found that localization was good for those clicks that approximately coincided, the solid black ones in the figure. The images of these clicks seemed to move slowly from one side to the other because of the small excess of pulse rate on the right side. There was no difference between hearing and vibratory sensation in the ability to inhibit, at least for a short time, the pulses on the left indicated by the lighter lines. But as a matter of fact it was possible to shift the attention from the pulses involved in coincidence to the other pulses that appeared only on the left side.

After this experiment was completed I had the impression that it was possible to go further and use a complete set of random clicks as shown in part *a* of Fig. 118. If it should happen that there are two or three clicks that appear at exactly the same moment on left and right sides, these clicks would produce a sensation in the middle, as illustrated in *b* of this figure. The other clicks on left or right with no counterpart will be localized on their own side. Under these conditions, with vibratory stimulation, there were three series of localized impressions, a left series, a right series, and a series in the middle representing the coincidences.

From these observations I concluded that the method would be useful for detecting periodicities of any sort in a complex temporal pattern. It seemed to be useful in certain physical systems as shown in Fig. 119. At *a* in this figure

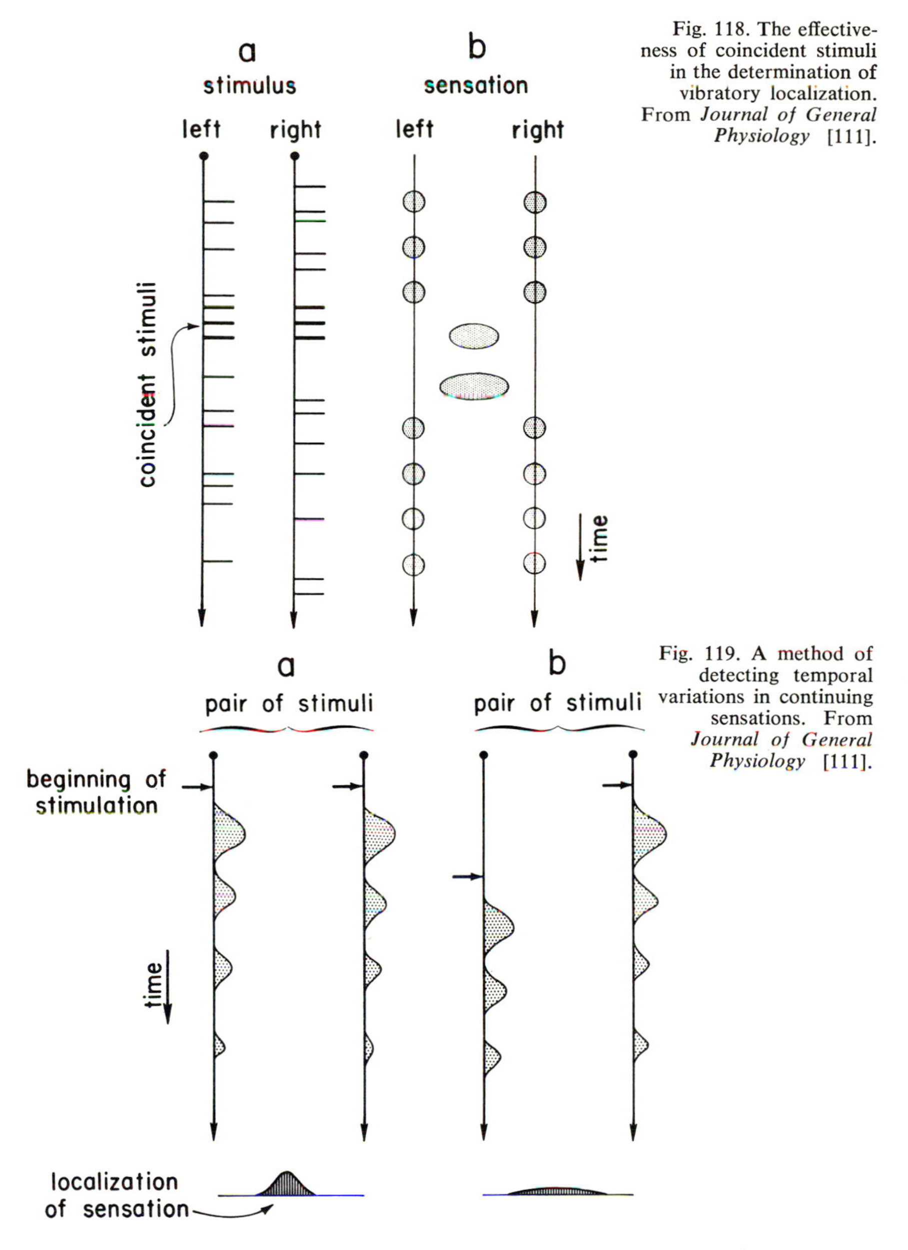

Fig. 118. The effectiveness of coincident stimuli in the determination of vibratory localization. From *Journal of General Physiology* [111].

Fig. 119. A method of detecting temporal variations in continuing sensations. From *Journal of General Physiology* [111].

are shown two stimuli that are identical in periodicity and temporal pattern. If such stimuli are presented simultaneously at two different points on the skin surface, there is a sharp localization halfway between the points. The length of the periodicity was determined by delaying one stimulator slowly and continuously as shown at *b*. At a certain time delay a sensation appeared again in the middle between the two stimulators, and this occurred when there was a coincidence between two maxima of the stimulus trains. When the delay was increased further this sensation in the middle moved to one side or the other and then disappeared, and there were two separate images, one under each stimulator. After further delay there was still another sensation in the middle, if the stimulus train was long enough. But if the periodicity pattern faded out like a damped vibration, then the sensations in the middle became more and more flat. The sharpness and magnitude of the middle images could sometimes be increased by varying the magnitude of the stimulus on one side so that the two maxima in coincidence for that particular time delay produced two sensory effects that were equal. In physiological systems the periodicity often is not constant, and also the amplitude may change as well as the length of the period, so that the sharpness of localization is impaired. Despite these difficulties the localization method is a good way of ascertaining whether a periodicity is present in complex stimuli.

I was mainly interested in the so-called slow reactions, as in the sense of taste. A question is whether it is possible to discover a periodic variation following the presentation of a constant taste stimulus. Certainly the neural processes are complicated, and the main problem is the periodic length of rhythmic variations, and whether their duration is of the order of seconds or milliseconds.

First an experiment was carried out with electrical stimulation of the tongue. As is well known, an electric current of the proper intensity produces a taste sensation at the stimulated area. If the area is large the quality is a combination of acid and salt with some side effects. To limit the area, the tongue was stimulated with two concentric electrodes 18 mm apart, and a series of direct pulses was applied for 10 milliseconds. The localization was perfect when the stimulus was introduced at the two electrodes at exactly the same time. By introducing a small time delay it was possible to move the sensation from the middle to the left or the right (Fig. 120). If the time delay was further increased for one stimulus (say, the left one) two separate sensations usually appeared, one under each of the concentric electrodes. For a time delay of about 3 milliseconds the sensation again became single and localized in the middle, and with a further increase in the time delay it moved to the right. This cycle of displacements of the sensation could be repeated several times as the time delay of one stimulus was continually increased. As Fig. 120 indicates, the length of each of these cycles was about 3 milliseconds. As is well known, the magnitude of taste sensation increases rapidly as the stimulating current is raised, and it is important to adjust the magnitudes for optimal localization.

The same experiments were repeated later

Fig. 120. Effect of progressive delay between successive pairs of electrically produced taste stimuli on localization of the sensations. From *Journal of General Physiology* [111].

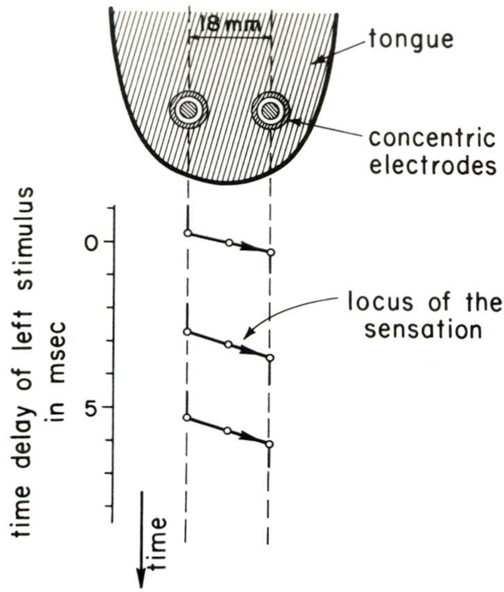

with chemical stimuli by changing the test solution applied to the tongue from tap water to a solution of salt, acid, sugar, or quinine for about a second, and then going back to tap water. Here too it was found that the constant stimulation of the tongue produced a taste sensation that was not constant like the stimulus but showed rhythmic variations as indicated in Fig. 121. Along with these variations there were changes in the lateral spread of the sensation, as indicated on the right of this figure.

The course of some sensory nerve pathways

For directional hearing it is reasonable to assume that there must be interconnections between the two ears to produce an interaction through which a time delay can bring about an inhibition. About three decades ago I spent a

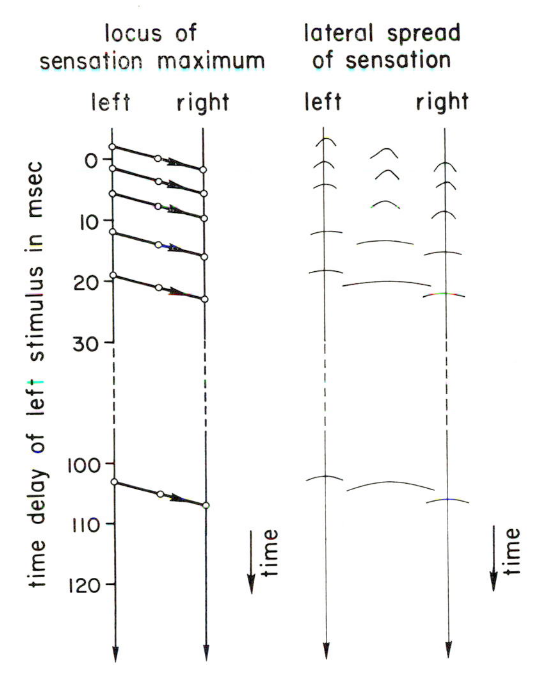

Fig. 121. Effect of progressive delay between successive pairs of chemical stimuli on taste localization. From *Journal of General Physiology* [111].

considerable amount of time tracing some nerve fibers from one hemisphere of the brain to the other to discover what place would be the most likely one for interaction between left and right ears. After having looked through the microscopic slides and records of mental hospitals and consulting the literature, I discontinued this attempt because I was unable to find any reasonable method of identifying this particular portion of the brain.

Later I was interested in the pathways between the two hemispheres not only for auditory tracts but also in relation to the skin sensations and their tracts. The thought was that perhaps the sensory tracts for the skin are simpler and easier to follow than the auditory ones. If so, I could learn something about the switching techniques used by the brain in the interactions of a two-point stimulus. Before going into anatomical work I wished to make sure that there was a similarity in this respect between vibratory and auditory sensations. From results already presented in these lectures it is clear that a similarity is present. Nevertheless I set up the apparatus shown in Fig. 122, which consists of two microphones with their two amplifiers leading to two vibrators adapted for skin stimulation. The microphones were placed at a distance of about 20 cm from a small loudspeaker that produced a series of clicks. These clicks were easy to localize acoustically. Then, however, auditory perception was excluded by the use of ear plugs and pads over the ears, so that sound localization was no longer possible. Yet the vibratory sensation was still easily felt on the chest. This vibratory sensation was localized in the middle of the space between the points of application on the skin when the loudspeaker was in the plane midway between the two microphones, but moved to the left or the right when the loudspeaker was moved correspondingly in the room.

This experiment suggests that a neural pathway should exist for vibratory sensations running from each side of the body to both hemispheres, just as for the ear. But this suggestion is not correct, and I do not believe it to be an easy matter to discover the correlates of the neural pathway for this kind of localization. The difficulty is discovered on displacing both of the vibrators to one side. When this is done there is absolutely no deterioration of the localization.

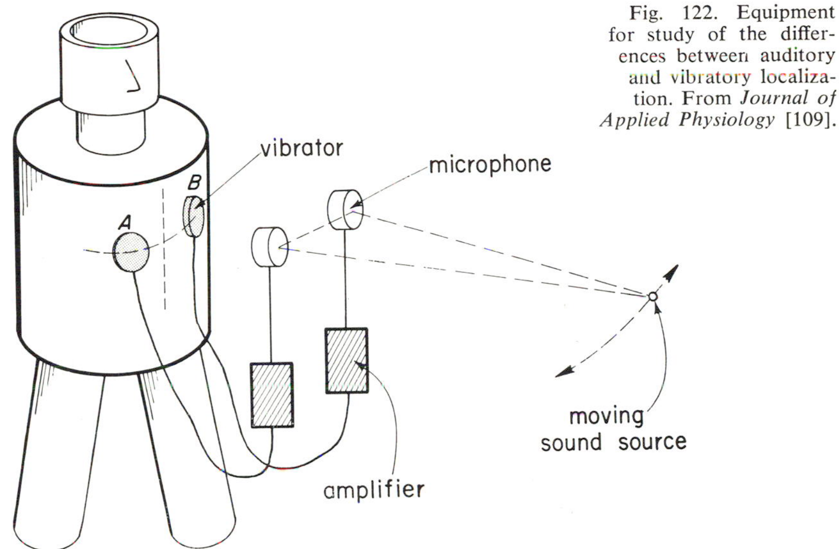

Fig. 122. Equipment for study of the differences between auditory and vibratory localization. From *Journal of Applied Physiology* [109].

It is possible to go farther and to place the two vibrators (*A* and *B* of Fig. 122) in a vertical position, one on the lower portion of the chest and the other on the upper portion. Even under this condition it is possible to cause the sensation to move from one vibrator to the other by introducing a time delay. From this evidence I conclude that localization both for directional hearing and for skin sensation is not restricted to an interaction between the two hemispheres.

The speed of nerve conduction

The speed of conduction in nerves is an interesting chapter in the history of physiology. More than a hundred years ago it was generally accepted that because this conduction is electrical in character it must proceed with the speed of electricity, which is 300 kilometers per second and therefore completely out of the range of measurements of that time.

It was Helmholtz (1850, 1852) who completely modified this mythical concept by showing experimentally that at normal temperatures the speed of nerve transmission in the frog varied between 25 and 40 meters per second. He measured the speed of sensory nerve transmission in man also and found it to be 60 meters per second. At the time this was a very important achievement, because even at that range of velocities special equipment had to be constructed, such as the contact pendulum, to produce the stimuli with sufficient accuracy. I have always been impressed by Helmholtz's achievement.

Later Taylor (1941) and Erlanger and Gasser (1937) showed that the speed of nerve transmission depends in high degree on the thickness of the nerve fiber, and the range of variation is great. Kohlrausch (1866) had reported speeds for man up to 225 meters per second.

Nowadays the speed of nerve conduction is regarded as analogous to the speed of conduction of a concentric electric cable; in such a cable the speed is determined by the capacity of the insulating layer between the two conductors and the resistance and conductance of the metallic conductors. For this analogy we prefer lower speeds, but Taylor (1941) and von Muralt (1958) have pointed out that the speed of nerve conduction is not a simple problem. For example, nerve fibers in the squid that transmit at a speed of 25 meters per second have a diameter 300 times larger than fibers in the cat that transmit at the same speed, as shown in Fig. 123. This fact makes it difficult to extrapolate from one nerve fiber to another.

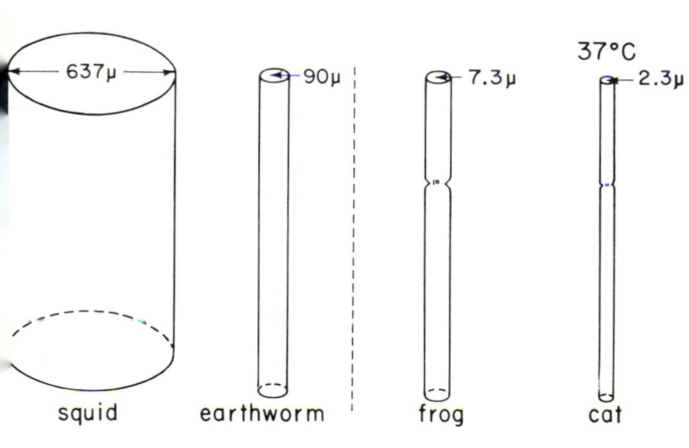

Fig. 123. Nerve fibers with the same speed of conduction but different diameters. After von Muralt (*Neue Ergebnisse der Nervenphysiologie*, 1958, Springer-Verlag).

It is possible that in different fibers the process of transmission is entirely different.

Because in many measurements of the speed of nerve transmission there is little control of the blood supply, and often anesthetics are used, I wished to repeat the method used by Helmholtz. The progress of a century made it possible to improve his technique. Today it is no problem to make the sensory magnitude for two stimuli exactly the same by the use of electrodynamically driven vibrators and attenuation boxes. The pulse can be shaped according to the frequency transmitted, as desired. Moreover, there is the possibility of repeating the stimuli in a rapid series, and it is easier to make observations with a series of clicks than with a single click. By the use of special vibrators, the stimulation can be confined to the skin surface, without involvement of deeper layers. Precautions can be taken also to avoid transmission of the vibrations on the arm to the bones where propagation occurs at speeds greater than in the nerve fibers. Hence a number of side effects can be avoided.

In his experiments on man, Helmholtz used two stimulators on the surface of the arm and found the time delay necessary to make the stimulus applied at the place closer to the brain seem coincident with the one applied farther away. A feeling of coincidence really represents a temporal coincidence of the maximum intensities of the two sensations. As noted earlier, the magnitude of sensation is the slower of two neural processes arising from stimulation. Hence there was a possibility that by observing the discharges at onset and bringing them to coincidence a greater speed might be obtained for this part of the discharge. To measure the onset portion of the discharge it is necessary to utilize the phenomena of localization. The localization phenomena produced by time delays were first described by von Hornbostel in 1923, and were not known to Helmholtz. I believe that there is a difference between observations of coincidence and observations of localization. Coincidence data give lower speeds as a rule, and localization effects can be used with high precision, as has already been shown. The difficulty with the localization experiments is that as the distance between the two vibrators increases the fusion of the two clicks to one becomes more and more difficult. However, this situation can be improved by enlarging the stimulation area and matching the clicks not only for sensation magnitude but also for timbre.

For precise observation of localization it is of utmost importance to adjust the magnitudes of sensations to be equal for both the stimuli used.

To assist this adjustment I chose two points on the arm (Fig. 124) with nearly the same skin structure and sensitivity. Then for these two points the stimuli were adjusted to give equal magnitudes of sensation by presenting these separately. Next an effort was made to obtain a sensation localized midway between the two stimulated regions by giving one vibrator (No. 2 in Fig. 124) a time delay of at least 1.2 milliseconds. By increasing or decreasing the time delay from this new zero point it was possible to move the vibratory sensation from one vibrator to the other just as was done earlier on the chest (horizontal position) and between two fingertips. These times are shown in Fig. 124 by the solid line (marked 0 db). Here "0 db" means that the two magnitudes of sensation

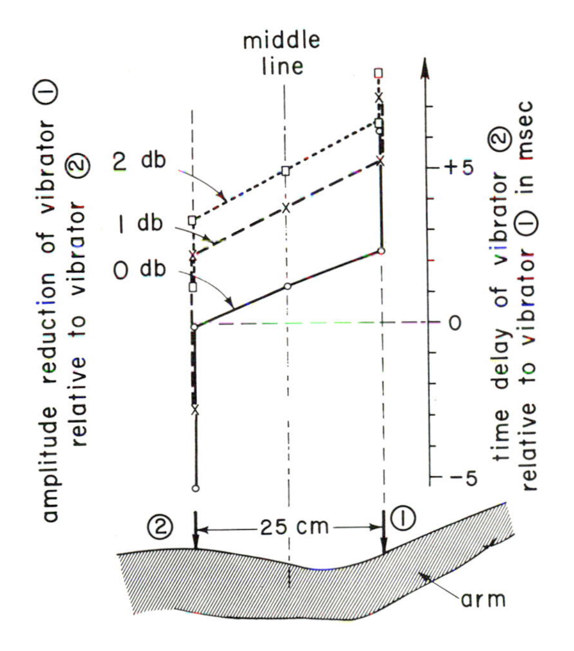

Fig. 124. The use of vibratory localization as a method of determining the speed of nerve conduction.

were made exactly equal. If vibrator No. 1 is made 1 to 2 db weaker, it is necessary to delay No. 2 by 5 to 6.5 milliseconds to cause the vibratory sensation to appear in the middle. This fact shows the importance of a correct adjustment of sensory magnitudes. Naturally a bracketing method was used. The speed obtained for precisely matched stimuli was of the order of 25 cm per 1.2 milliseconds or 208 meters per second. This speed is higher than usual but it is the speed of the nerve spikes representing the onset of the stimulus and used in localization.

There is always a chance that there is a transient delay during transmission from the tip of the vibrator to the receptor that might have an effect on the speed. Measurements made on other parts of the body indicate that such an effect is improbable, but to make sure the experiment was repeated with electrical stimulation by means of the concentric electrodes shown in Fig. 125. With a proper adjustment of current magnitude the result will be a pressure sensation and not an electric shock as at higher intensities. The only drawback with electric stimulation is that matching for equal magnitudes of sensation is more difficult than for vibratory stimulation with clicks. The results for the speed of nerve transmission by this method are shown in Fig. 126.

The measurements were repeated with stimulation along the entire body from neck to toe, and about the same speed was obtained. Of course there may be many objections to the calculation of the speed of nerve transmission from localization data, especially on the basis of the fact that the skin is not uniform. This is true, but observations were made with one vi-

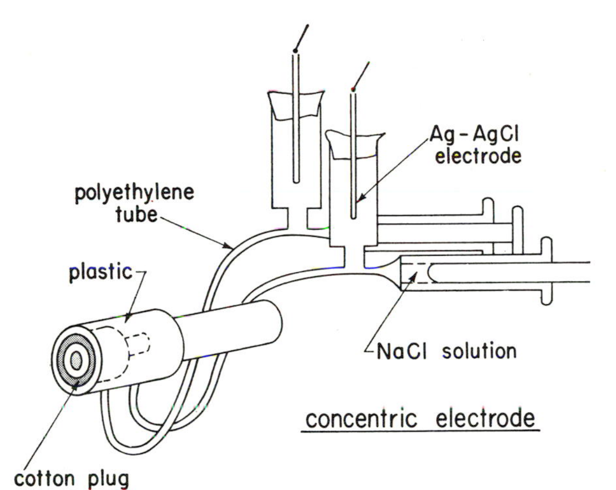

Fig. 125. A concentric non-polarizable electrode for electrical stimulation of the skin. From *Journal of Applied Physiology* [109].

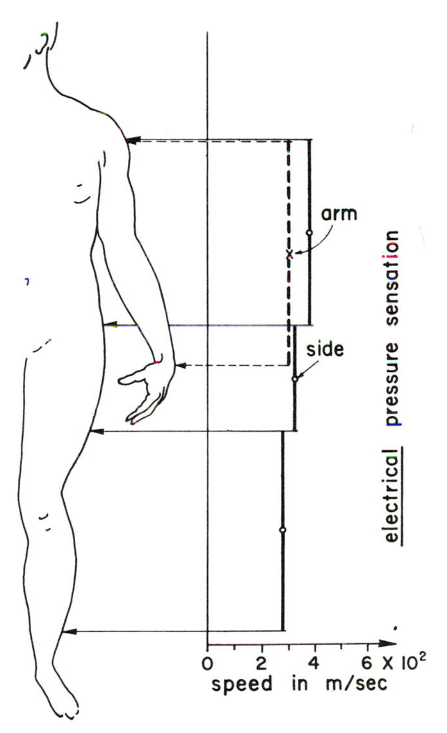

Fig. 126. Speeds of nerve conduction over the body, as measured by the localization method with electrical stimulation. From *Journal of Applied Physiology* [109].

brator on the outside of the arm and the other on the inside, and also from the outside and inside of the thigh, which are regions where the skin varies considerably in thickness, and the results were the same. A major objection might be that the results do not agree with the earlier ones on nerve transmission, but as has already been pointed out the earlier ones were based on a coincidence of sensory magnitudes whereas these depend on the phenomena of localization. It is possible with the localization methods to show that the speeds change with warming and cooling of the arm. It is even possible to show that an electric shock applied 10 seconds before the observations causes a reduction in the speed. This may be a result of vascular constriction or of some other, unknown phenomenon. But, generally speaking, any kind of disturbance reduces the speed, and it is not surprising to find low speeds reported in the literature.

As found earlier, a time delay between stimuli presented to the two fingertips had to be only 1 millisecond to displace the sensation from one finger to the other. As may be seen in Fig. 127, if the two index fingers are brought into contact two sensations are felt that are equal in area and localized near the contacting surface. When the forefinger touches the lip or cheek repeatedly with brief, light touches, the impression is that the pressure sensation is localized on lip or cheek and not on the forefinger. The reason probably is that the nerve discharges from lip or cheek arrive at the perceptive center sooner than those from the fingertips. This time delay exceeds 1 millisecond.

This situation can be reversed by touching the leg with the fingertip. Then the fingertip sen-

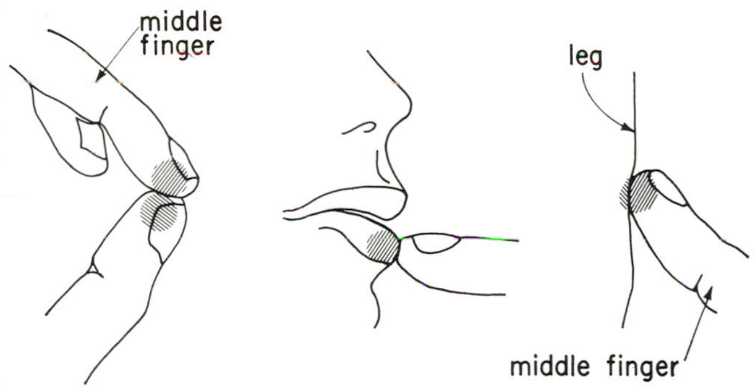

Fig. 127. Localization of touch sensations under different conditions. From *Journal of Applied Physiology* [109].

sation arrives at the brain first and will inhibit the sensation from the leg, and the touch is felt mainly in the fingertip. This experiment can be repeated along the whole body, and the results give a distribution of sensations as shown in Fig. 128. There is a difference when the sole of the foot is touched because of the great variations of thickness of the skin there. Otherwise

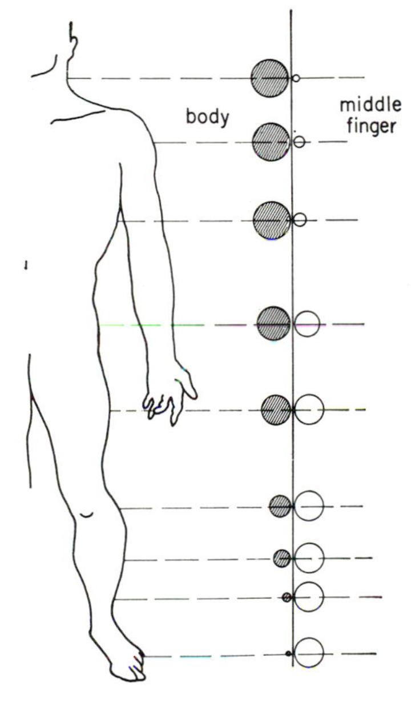

Fig. 128. The perceived areas of sensation in different bodily regions touched with the fingertip. From *Journal of Applied Physiology* [109].

it is found that at the middle of the thigh, where the distance from the brain is about the same as for the index finger, it does not matter whether the outer side or the inner side is touched.

It is known from electrophysiological studies that the nerve fibers for taste are small in diameter, and therefore their speed should be low relative to that for vibratory sensations. It is therefore of special interest to observe the speed of taste fibers on the tongue. For this purpose a plastic plate was used as in Fig. 129. Here only the openings in the plate are shown. The switching devices are on the other side, and are the same as shown earlier in Fig. 83. If the fluids in openings *A* and *B* are changed simultaneously

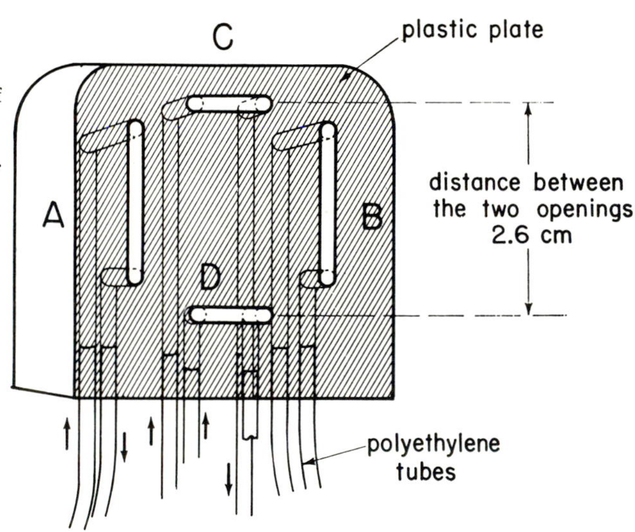

Fig. 129. A plastic plate containing four fluid channels, used in the measurement of speed of nerve conduction for taste. From *Journal of Applied Physiology* [109].

for about 1 second, the taste sensation is localized midway between them as described earlier. Openings *C* and *D* were used for the speed measurements. It was evident at once that when there was no time delay between the two stimuli the sensation was always localized at *C*, which was at the back of the tongue. Therefore it was necessary to deliver stimulus *D* somewhat earlier in order to position the stimulus midway between the two openings. When the test solution and the water were at 23° C the time difference was 0.7 milliseconds ± 0.2 milliseconds. The distance between *C* and *D* was 2.6 cm, hence the mean speed of transmission at this temperature was about 40 meters per second. When the concentrations of solutions of sucrose, salt, and quinine were adjusted so that they gave a magnitude of sensation equal to that of a 0.01 per cent solution of hydrochloric acid, there were no striking differences among the results obtained with these solutions. When the solutions were warmed to 38° C, however, the mean value of the transmission speed rose to 80 meters per second. Again the problem of controlling the temperature was so difficult that no attempt was made to improve the precision of the equipment or to use a large number of observers.

In this discussion of the use of localization phenomena to determine the speed of nerve transmission it was always assumed that the neural time pattern produced by the two stimuli is conducted to a higher level and that the localization occurs at that level. The evidence favoring this assumption is that displacements of the stimuli in a transverse direction anywhere on the body did not produce time delays.

Some discrepancies between psychological and electrophysiological observations

Progress of the last decade in the field of physiology is due mainly to electrophysiology. This is a new tool that has made it possible to record the electrical activity of single units and to discover what information these units transmit to higher neural levels. At present there seem to be limitations to the interpretation of electrophysiological data. Dijkgraaf (1963) pointed out in a series of papers that electrophysiological and behavioral methods for the study of sensory functions must lead to different results. The reason is that in a complex system of transmission lines going directly to higher levels, with inhibitory and feedback connections, it is difficult to discover which feature in the complex produces the sensation. For example, it is not possible at present to decide what is the electrophysiological equivalent of the magnitude of sensation, though psychologically this magnitude is a simple and well-defined attribute.

I have gained the impression that the onset of a nerve discharge is used for localization, whereas the later discharges serve for recognition of the sensory magnitude. This situation clearly indicates the difficulty of using electrophysiological data alone in the study of neural activity. There is no secure electrophysiological evidence to show that the duration of a sensation enters into the determination of its perceived magnitude. A more general question is what part of the temporal pattern of a nerve spike is utilized by man and other animals in the determination of behavioral reactions.

A difficulty comes from the presence of spontaneous activity, as described by Hoagland

(1932) and Burkhardt (1961). This activity may be of importance in the behavior of certain animals. It is to be expected if there are sense organs that operate on the principle of compensation, such as the vestibular organs. With such organs it is the change in the spontaneous activity that is of importance. If there is a change in the spontaneous activity of two mutually compensating systems it may be a result of adaptation. Such a change is unimportant, and I find it difficult to distinguish clearly between the change that is important in stimulation and the one that is only adaptive.

A second problem in electrophysiology is that the use of the oscillograph places an emphasis on the temporal pattern of nerve activity recorded at a particular point. For studies of single neural units the temporal pattern is expressed mainly in variations of the rate of spike discharges. Yet in the process of recording spike discharges it is likely that adaptation of the receptor at a higher level is already included. This information may not be important in our effort to discover how the nervous system operates. In general I have found that the reading and memorizing of oscillographic records of temporal patterns is exceedingly difficult.

It is obvious that the spatial pattern of neural processes is highly important, but unfortunately it is difficult to implant several electrodes in such a way as to give a clear indication of the spatial variations of neural activities. It will take time to improve our techniques of recording in this regard. The successful recording with a single microelectrode gives no guarantee that we are recording the activity of a single unit, for a group of units may discharge in synchrony.

There is a possibility also that in adjacent nerve tracts there is interaction producing inhibition like that shown in Fig. 36. As shown there, a small excess of amplitude seems to be sufficient to synchronize all neighboring organs to a common frequency and phase.

The most successful stimulus in electrophysiology is the click. It is a combination of an on and off effect. Peculiarly enough the effects of sound that may be detected by recording the impulses of the higher auditory centers are much less obvious for a continuous tone than for a click or a tone pulse. This is contrary to our psychological observations, in which we find continuous steady tones to be quite as effective as tone pulses. This is a problem for which I do not at the present see any indication of a solution. Our main hope at the moment is that the mathematical study of possible neural networks will bring to attention some neural phenomenon not yet observed in electrophysiological experiments. I only hope that these network studies do not develop into a science resting on its own axioms and thereby becoming a sort of non-Euclidian network neurology.

These problems needed to be mentioned because it is useful to recognize that the most promising development of the present time is the concurrent application of electrophysiological and psychological methods.

Similarities between psychology and electrophysiology

The precision shown in the localization of a sensation during a variation in the time relations between stimuli raises questions that at the moment are exceedingly difficult to answer.

Every stimulus onset produces a large series of neural discharges, and these discharges do not seem to be as precisely related to the stimulus onset as the exactness of localization requires. Probably the onset activity is simplified by inhibitory processes so as to make the nerve transmission more precise. Also a large number of end organs are usually stimulated at the same time, and together they contribute to the formation of a relatively stable mean. Without inhibition of the unimportant nerve discharges it is difficult to conceive how a rotating tone becomes so well defined and free of sudden variations.

It is clear that there is a disparity between psychologically obtained precision and our electrophysiological observations. We need to consider how crucial this discrepancy is. Because in electrophysiology the animals are under abnormal conditions of anesthesia, some degree of hypoxia, and the like, it would be helpful to observe nerve discharges directly in a human subject. It is probably possible to do so for vibratory stimulation at low frequencies. In electrophysiological studies on cats stimulated with sinusoidal vibrations at low amplitudes, Keidel (1956) showed that near the threshold there was a neural discharge at every second or third wave, recorded either from the skin or vibrissae. This kind of discharge is shown schematically in *a* of Fig. 130. When the amplitude of vibration was increased the discharge changed as shown in *b*, with a discharge at every wave; for still higher amplitudes there was an irregular pattern as in *c*, with from 1 to 3 discharges for each wave. This irregularity continued for all the larger amplitudes.

Fig. 130. Schematic representation of the discharges of nerve fibers for different amplitudes of vibratory stimulation. Redrawn from *Journal of the Acoustical Society of America* [85].

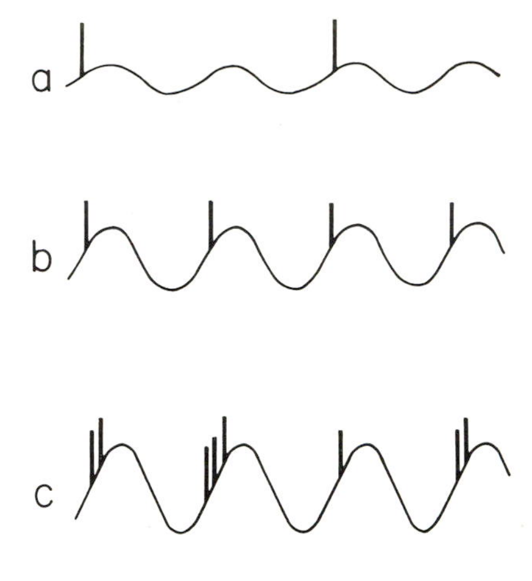

If the moment of the nerve discharge is recorded relative to the phase of sinusoidal vibrations it is soon seen that the discharge does not occur at exactly the same point in each wave. It will sometimes slip a little ahead, and sometimes be delayed. Anesthesia and other conditions may cause large variations in the moment of onset.

It was found that the increase in the rate of discharge with increasing amplitude can be recognized in some degree for slow vibrations applied to the fingertips. With vibrations of 10 per second, a gradual increase in amplitude from below the threshold to about 10 db above it gives the impression that the vibratory pitch of the sensation increases also. A rough estimate of this "pitch" can be made, as shown in Fig. 131. The vibratory octave can be defined by comparing the sensation for 5 cps with that for 10 cps, at the same magnitude of sensation. After a few weeks of training the "pitch" change of one octave can be remembered very well. The

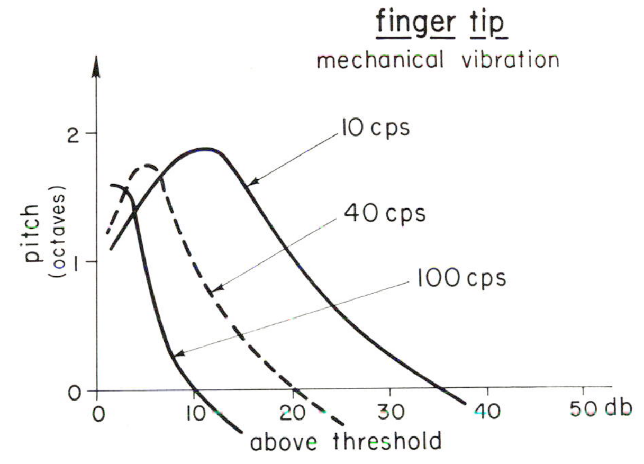

Fig. 131. Effect of the intensity of mechanical vibration on vibratory "pitch." From *Annals of Otology, Rhinology, and Laryngology* [102].

unexpected feature of the experiments recorded in Fig. 131 is that during the increase of vibratory amplitude the "pitch" suddenly began to fall, and fell rapidly through two or perhaps three octaves. The drop begins earlier for high frequencies such as 100 cps. The decline in "pitch" is mainly a result of the fact that the smoothness of the vibratory sensation ceases at a certain amplitude, and the sensation then becomes modulated with a periodicity that is far below the periodicity of the stimulus frequency.

The manner in which the smoothness of the vibratory sensation changes during an increase in the amplitude of vibration is shown in Fig. 132. Not every observer describes the experience in the same way, but always the sensation is smooth for the highest "pitch," as shown at "3" in the figure. For small amplitudes, in general, there is present a sort of irregular modulation. This psychological observation indicated that the nerve discharges were not able to follow the amplitude of vibration completely, and the discharges became irregular and out of

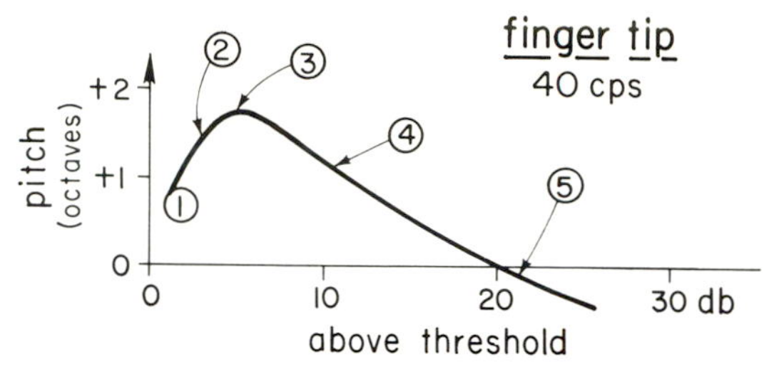

Fig. 132. Representation of perceived vibratory roughness at different levels above threshold. From *Annals of Otology, Rhinology, and Laryngology* [102].

phase. I had the impression that this irregularity was typical for mechanical stimulation and therefore the experiment was repeated with electrical stimulation. The same results were obtained. With an electric stimulus it is easy to produce a sensation of "pitch" at a certain sensory magnitude by presenting a series of short clicks, and to demonstrate that these clicks cannot be followed by the neural discharge. Some results are presented in Fig. 133. The form of the clicks is shown at the top of the drawing.

When the number of clicks per second was low and this rate was suddenly doubled (as from 200 to 400 per second), an observer trained to recognize an octave variation reported that the "pitch" was doubled. Peculiarly

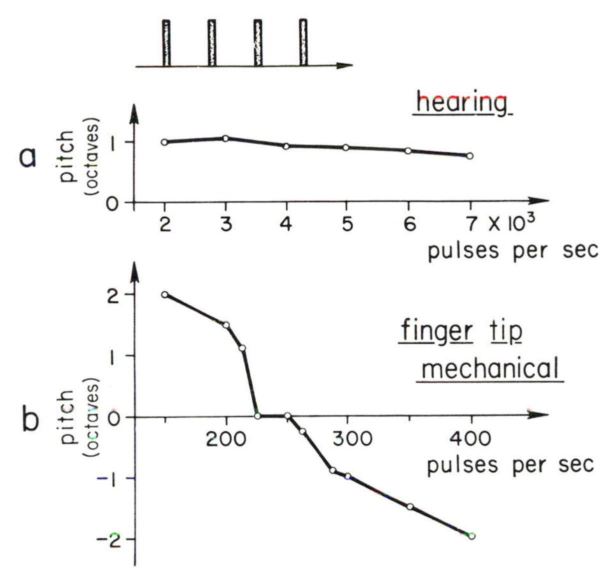

Fig. 133. A comparison of pitch changes as a function of the doubling of frequency, for hearing and vibratory stimulation.

enough, however, this doubling did not occur when the frequency was changed from 300 to 600 per second; then the observer failed to note any "pitch" change. When the frequency was changed from 400 to 800 per second, the "pitch" definitely fell. This evidence shows that the nerve discharges cannot follow the increase in stimulus frequency at high rates. At around 300 pulses per second there occurs a sort of demultiplication of the neural process, a method widely used in the nervous system to transmit information and to avoid unwanted feedback loops that might cause oscillations. The interesting thing is that electrophysiological recording shows this demultiplication very well. Figure 134 shows records taken by Rose and Mountcastle (1959). There is perfect agreement between mechanical "pitch" sensations on the fingertip and electrical recording.

Because the agreement is so good, and there is a relation between the phase of the vibrations and the nerve discharges, we have the question of what happens when we touch the surface of the skin with the tip of a vibrator and there is not just one point vibrating in phase with the tip but a whole traveling wave with a whole series of phase relations (Fig. 13). If a close relation is to exist between the phase of the vibrating tip and the nerve discharges under these conditions, it will be necessary to assume that all vibrations that are not in the immediate vicinity of the tip are inhibited and do not contribute to the sensation of "pitch." This assumption is in agreement with the observation that the vibratory sensations are localized exactly below the vibrating tip and we do not perceive the large area over which traveling waves move along the surface (as can be proved by stroboscopic illumination). Because the traveling waves have a great speed, it is necessary to conclude that the inhibition is rapid and in the millisecond range. Otherwise the large stimulated area would be perceived at the onset of the vibration.

Returning now to the subject of inhibition, it may be pointed out that for the explanation of the Mach bands it is important that the concept of the neural unit be demonstrated at the higher levels of the nervous system by the use of microelectrodes, and that the indications of these electrodes be consistent with the psychological observations. Mountcastle and Powell (1959) found that on stimulating the skin and

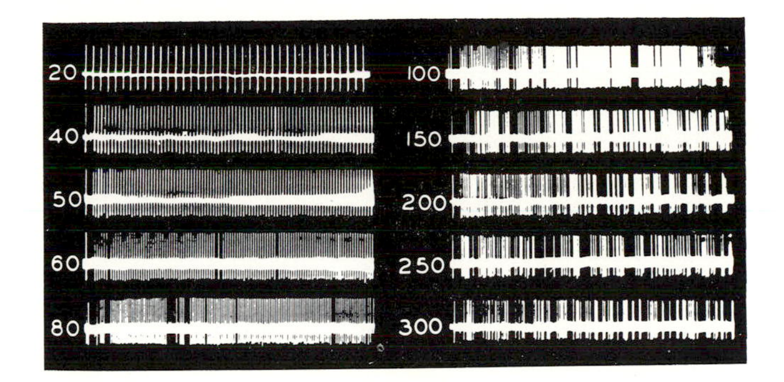

recording from places in the dorsal column, in the thalamus, and in the cortex, it is possible to map out areas of excitation that are surrounded by areas of inhibition, as shown in Fig. 135. These results are in close agreement with the observations described earlier for stimulation of the palm. Likewise in the monkey they

Fig. 134. Discharge patterns for cutaneous nerve fibers under vibratory stimulation at various rates from 20 to 300 per second. At the higher rates, from 80 per second on, the fibers failed to maintain a continuity of action. From Rose and Mountcastle (*Handbook of Physiology*, I. *Neurophysiology*, 1959, 1, p. 409).

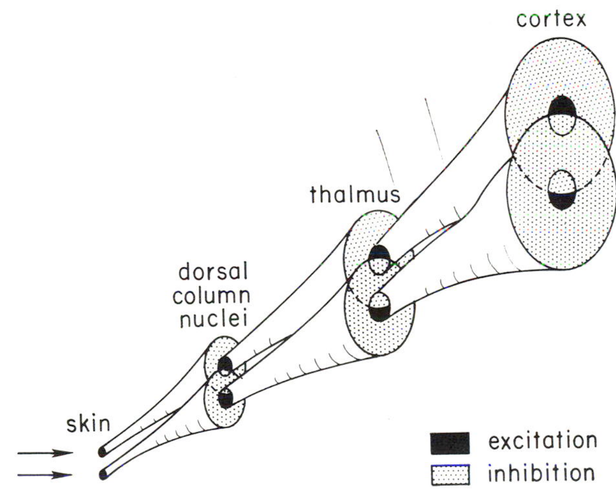

Fig. 135. The effects of point stimulation of the skin as recorded at successive levels of the nervous system. From Mountcastle and Powell (*Bulletin of the Johns Hopkins Hospital*, 1959, 105, pp. 201-232).

found a sensation area surrounded by an inhibitory area as represented in Fig. 136. More recently, Suga (1965) showed that in the auditory system of the bat every response area is surrounded on both sides by an inhibitory field. These results are given in Fig. 137. Here the excitatory area is represented by open circles and the inhibiting area is shaded. The recording was done with microelectrodes in the inferior colliculus during stimulation by tone pulses.

After it became clear that lateral inhibition is readily demonstrated in the higher portions of the nervous system, there remained the question whether it first arises in the end organs. There is little electrophysiological evidence on most of the sense organs except vision. Kohata (1957) indicated that pressure exerted on the eyeball reduces the inhibition in the eye. This evidence indicates that inhibition is largely present at the retina.

The best evidence, however, comes from the observations of Ratliff and Hartline (1959) on the eye of *Limulus*. They used a screen in front of the eye with an opening so small that light could stimulate only a single ommatidium. Then the ommatidium acted like a simple photocell and the rate of nerve impulses represented closely the intensity of the light acting on the unit. No inhibition was present because other ommatidia were not illuminated. If the light beam was made to move across the ommatidium the discharge rate rose suddenly. This sudden change in the discharge rate is shown in Fig. 138 by the curve with triangles.

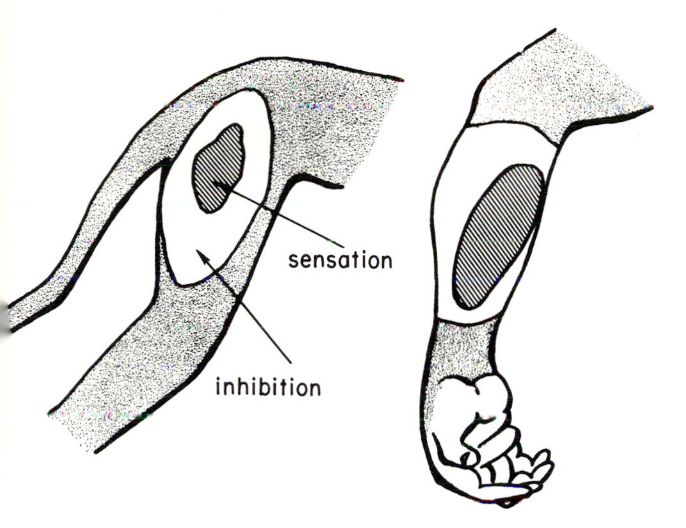

Fig. 136. Patterns of reinforcement and inhibition for skin stimulation, observed in the thalamic nucleus complex of the monkey. From Mountcastle and Powell (*Bulletin of the Johns Hopkins Hospital*, 1959, 105, p. 221).

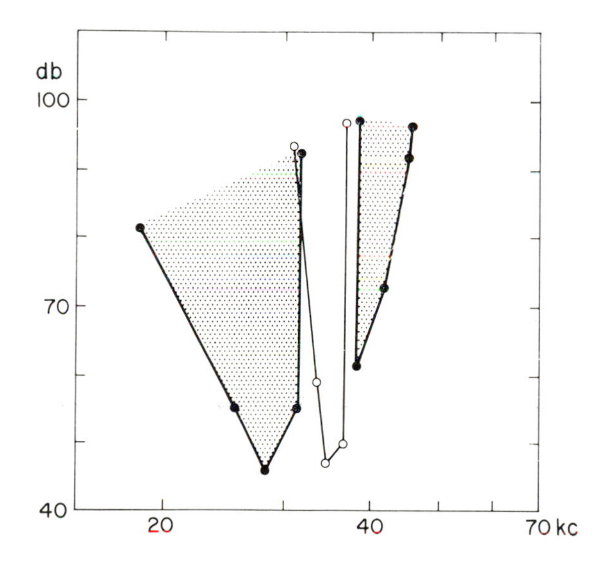

Fig. 137. Patterns of inhibition (dotted areas) as observed with a microelectrode in the inferior colliculus of bats, for tones of various frequencies. From Suga (*Journal of Physiology*, 1965, 179, p. 30).

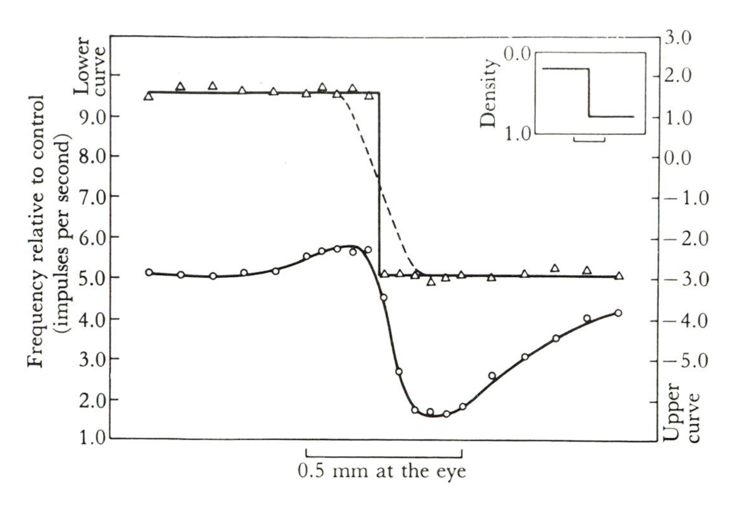

Fig. 138. Changes in the discharge rate of a single ommatidium on stimulating with a contour between light and dark, when the contour was moved in small steps across the ommatidium, shown under two conditions: when only this ommatidium was exposed to the light (curve with triangles) and when other neighboring ommatidia were stimulated also (curve with circles). From Ratliff and Hartline (*Journal of General Physiology*, 1959, 42, 1241-1255).

When, however, the screen was removed and the discharges of the same ommatidium were recorded for the same lateral displacement of the light beam, the lower curve of this figure was obtained. The ommatidia around the one from which the recording was made were now illuminated, and their action was to inhibit the the action of the central unit, producing an overshoot and an undershoot at onset and also at the cessation of stimulation, just as was observed in the experiments already described with the Mach bands. These effects are so large as to suggest that a large part of the lateral inhibition that occurs in vision is already present in the end organ. There is evidence also that the effects of edge contrast and the phenomena shown in the Mach bands can be followed all the way to the cortex. Here the significant observations have been made by Jung (1959, 1961) and by Baumgartner and Hakas (1959).

V · FUNNELING AND INHIBITION IN HEARING

In hearing the physical stimulus is well defined. The pure tone is one of the most simple of physical stimuli and is fully specified in terms of its frequency, sound pressure, and phase. Under most conditions these three features can be specified at the entrance to the external auditory meatus.

If a tone is not pure it may be considered as the sum of a series of pure tones, according to Fourier's theorem. Ohm's law of hearing, which was the statement that the ear performs a Fourier analysis on a complex sound, extended this simplicity of the stimulus to the ear's performance.

Unfortunately the sound pressures at the external auditory meatus undergo some complications in their action upon the tympanic membrane of the ear. Moreover, as the action proceeds farther to the inner ear the vibrations of the footplate of the stapes are transformed into extremely complex spatial patterns of vibration that even today are poorly known. Hence we are in difficulty in trying to determine the forces that stimulate the hair cells. This transformation of the simple mechanical stimulus into a complex vibratory pattern is a necessary part of the rough mechanical frequency analysis that occurs in the cochlea. Pitch discrimination results from a peculiar interaction between the preliminary mechanical frequency analysis and further neural processing. The neural processing consists mainly of funneling and inhibition in the interaction between neighboring end organs, and is similar to the processes occurring in the retina and on the skin.

In a consideration of the funneling process it is necessary first to examine the anatomy of the inner ear.

The auditory mechanism

As indicated in Fig. 113 the footplate of the stapes is set into vibration by the eardrum and ossicles, and this vibration acts on the elongated canal of the inner ear, which is called a cochlea because of its spiral form (Fig. 139). In Fig. 113 this spiral canal was represented as a straight tube divided by a partition into two portions. The partition consists partly of bone and forms a kind of frame enclosing an elastic membrane, the basilar membrane. The canals on both sides of the basilar membrane are filled with a fluid called perilymph. When the stapes moves inward the fluid is pushed along the upper half of the canal and along the membrane until at a more or less well-defined place this membrane yields, and the fluid movements continue in an opposite direction along the canal to an opening called the round window. Always the volume displacement produced by the movement of the stapedial footplate is equal to the volume displacement involved in the bulging of the round window. As a safety valve there is an opening at the apical end of the cochlea, and a direct displacement of the stapedial footplate can cause a flow through this opening without a deformation of the basilar membrane.

An examination of the basilar membrane under stroboscopic illumination shows the sinus-

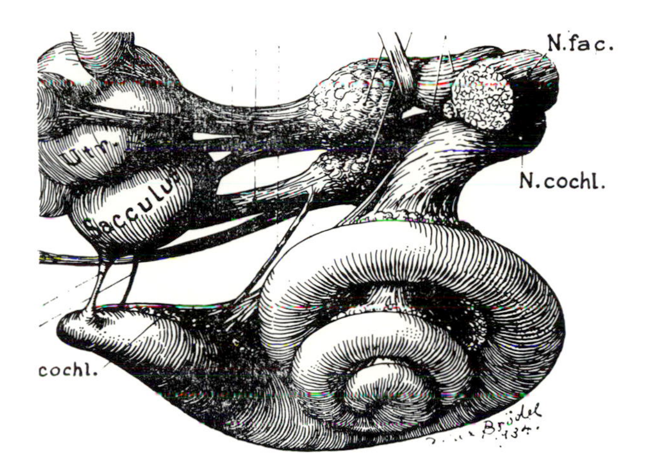

Fig. 139. Max Brödel's drawing of the membranous labyrinth and its nerve supply, in man. From Hardy (*Anatomical Record*, 1934, 59, 403-418).

oidal vibrations of the stapedial footplate and shows traveling waves running from the narrow end of the basilar membrane to the softer portions of the basilar membrane near the helicotrema.

Many years ago there was strong opposition to the assumption of traveling waves along the basilar membrane. However, a simple experiment will demonstrate that traveling waves are always present when a vibration is communicated to an elongated elastic band. As shown in Fig. 140, traveling waves may pass along a thread, an elastic band, a spring, or even a water surface. Whenever one of these systems makes contact with a vibrator source there are traveling waves that move away from this source. The pattern is just like that of the waves moving from the stapedial footplate to the apical end of the cochlea.

Comparative studies show that in the development of more complex members of the ani-

Fig. 140. Vibrations set up at one place on a flexible, continuous substance will produce traveling waves. From *Les Prix Nobel en 1961* [101].

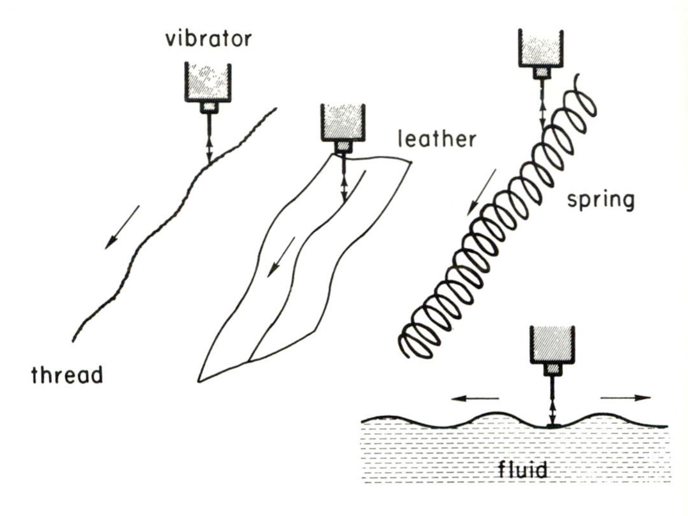

mal kingdom there was a continual increase in the length of the basilar membrane and in the number of hair cells. Figure 141 represents the development from the turtle to higher forms like the guinea pig and man. Hence there must be a relation between the length of the basilar membrane and the improved hearing of the higher animals. Because absolute sensitivity, in terms of the minimum detectable pressure

Fig. 141. Forms of the membranous labyrinth in the turtle, bird, and man. From *Akustische Zeitschrift* [45].

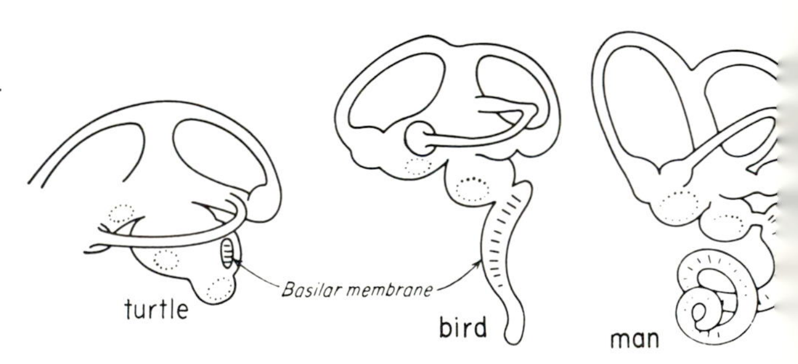

amplitude, is about the same for all animals, the increased length of the basilar membrane must be connected with the better pitch discrimination in higher animals, a discrimination that acquires great biological significance in the understanding of speech sounds.

The stiffness of the basilar membrane decreases constantly from its value near the stapes to a much smaller value near the helicotrema, and this variation is about the same, 100-fold, in a great variety of animals from pigeon to man. Under these conditions there is a maximum of the traveling wave that varies according to frequency. For high frequencies this maximum is near the stapedial footplate, whereas for low frequencies it is near the helicotrema, as shown in Fig. 142. This variation in location for different frequencies produces a rough sort of frequency analysis, in that high frequencies stimulate more strongly the parts near the basal end of the cochlea and low frequencies involve the parts near the helicotrema.

Figure 143 shows the instantaneous amplitudes produced along the entire basilar membrane for a tone of 200 cps (heavy solid and broken lines) and also shows the maximum displacements of the traveling wave in different regions of the basilar membrane (lighter broken lines). This drawing shows clearly that the amplitude maximum is flat. The problem of traveling waves in the cochlea is one of special interest because these waves occur along a surface that is continually changing in mechanical properties. Under this condition the waves are unique, and different from waves commonly encountered. The resulting phenomena arise from this inhomogeneity of the medium.

Fig. 142. Envelopes of vibration over the basilar membrane in man, for different tonal frequencies. The abscissa represents position along the basilar membrane measured from the stapes. From *Akustische Zeitschrift* [43].

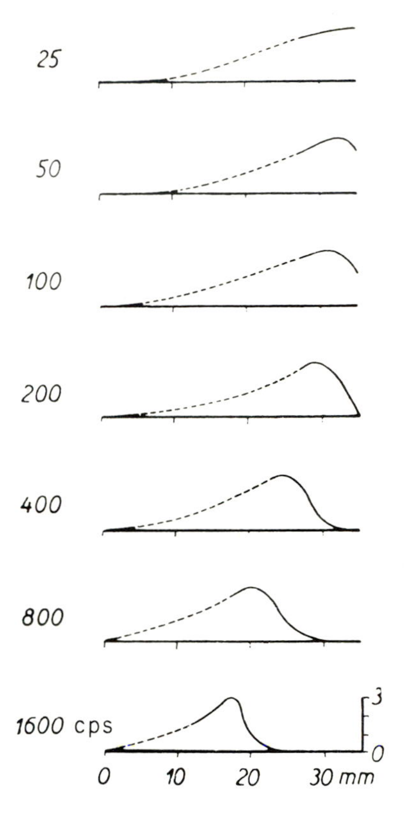

Fig. 143. A traveling wave along the human basilar membrane at two successive instants (heavy solid and broken lines), and the envelope formed by the positive and negative waves as they move up the cochlea (lighter broken lines). From *Journal of the Acoustical Society of America* [47]; see also [2].

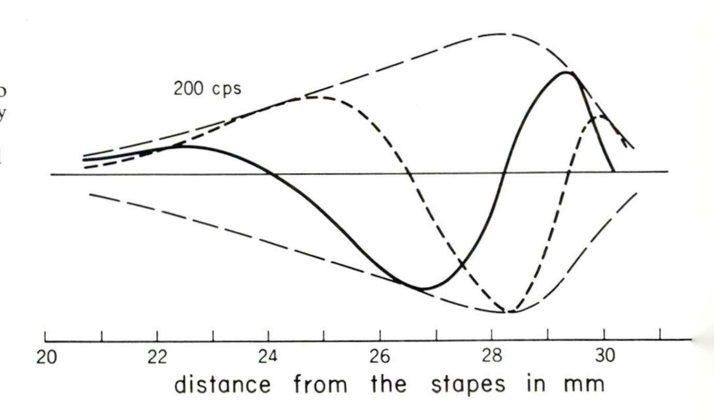

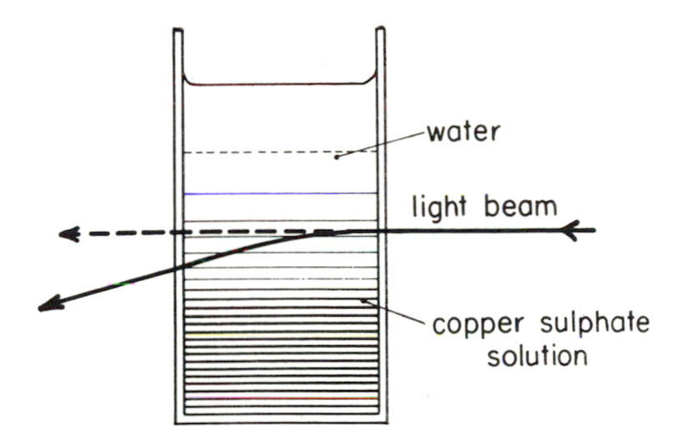

Fig. 144. Bending of a light beam at the boundary between two media. From *Proceedings of the National Academy of Sciences* [78].

Perhaps the simplest way of showing the abnormalities of wave propagation in an inhomogeneous medium is the experiment indicated in Fig. 144. A concentrated solution of copper sulfate is placed in a square trough. Above this solution, with the aid of a pipette, is slowly introduced a quantity of clear water. After about 10 minutes the blue copper sulfate solution will have diffused somewhat into the water to form a region where there is a continuous transition from dense copper sulfate solution to water. At first glance there seems to be no reason for a beam of light traveling horizontally through this optical system not to be propagated in a straight line, because the optical density along a particular horizontal line remains constant—the density is only changing in the vertical direction. Nevertheless, the usual optical law that light travels in a straight line fails to hold under these conditions. The beam bends in the manner indicated. This is a well-known experiment (Wiener, 1893).

The same behavior has been found for acoustic waves. They too show bending under certain conditions. When a sound wave passes over the heads of people in an audience, as shown in Fig. 145, the beam is bent toward the audience (Janovsky and Spandöck, 1937). Such bending produces an excess of damping of the waves.

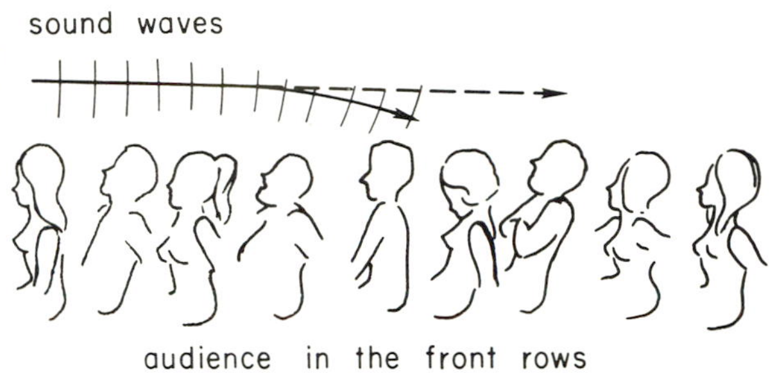

Fig. 145. Bending of sound waves passing over the heads of an audience. Redrawn from *Proceedings of the National Academy of Sciences* [78].

Physical and cochlear models

A model constructed for the purpose of indicating some of the peculiarities of traveling waves in inhomogeneous media is illustrated in the next few figures. An inhomogeneous medium is produced mechanically by the use of a series of pendulums of continuously varying lengths. In Fig. 146 the length is shown as increasing from right to left. Each pendulum was coupled to its neighbors by means of a small ball suspended between an adjacent pair of strings, and the weights of the balls were varied according to the level at which their strings were attached. The pendulums were driven by means of a tube rotating about its axis, and their energy came largely from the motion given to their points of suspension. The rotation of the tube could be given the desired periodicity and amplitude by the swinging of a heavy driving pendulum as shown.

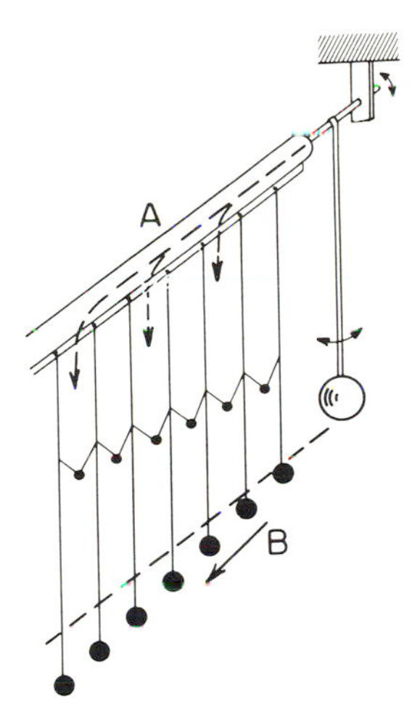

Fig. 146. Pendulum model of the basilar membrane. From *Proceedings of the National Academy of Sciences* [78].

When the system was set in action by the driving pendulum at the proper frequency there was a maximum excursion for a certain section of the series of pendulums and at the same time there was a traveling wave that moved from right to left as pictured in Fig. 147. The interesting feature of this type of wave is that the energy always moves from a stiffer section of the system to a softer one. For better illustration of this principle, a section of "soft" pendulums was displaced by a mechanical device, and this device suddenly withdrawn. As Fig. 148 shows, after withdrawal of the device there appeared only a very small traveling wave going toward the stiffer section to the right in the figure, while the greater part of the energy of the waves moved toward the softer sections. Therefore the longest pendulum was made to swing with an amplitude that was a multiple of the displacement produced at the beginning.

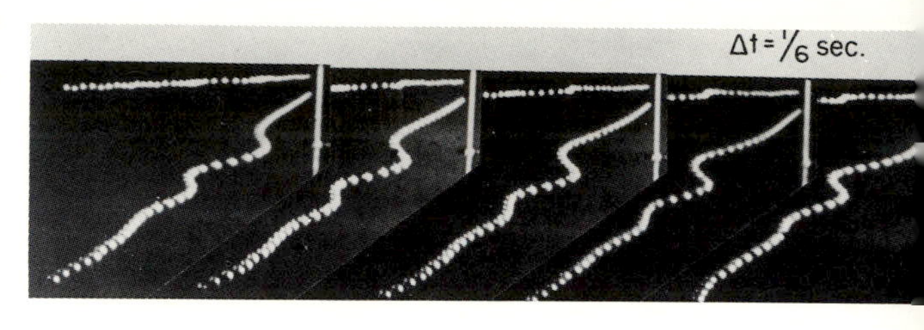

Fig. 147. Traveling waves along the pendulum model. From *Proceedings of the National Academy of Sciences* [78].

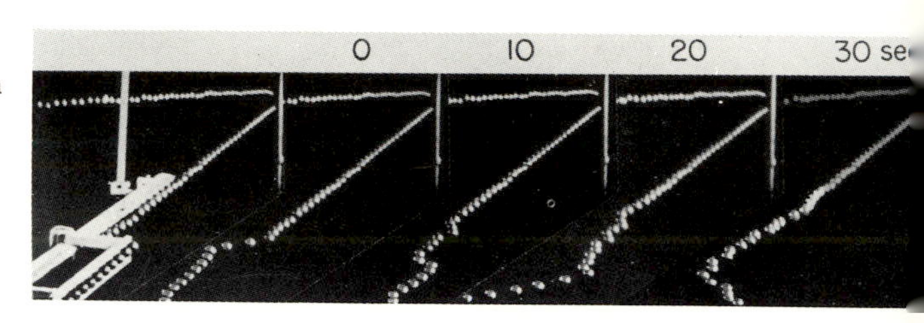

Fig. 148. Passage of traveling waves from stiff to soft pendulums. From *Proceedings of the National Academy of Sciences* [78].

This is a poor model, and it has only one feature in common with the cochlea, which is the presence of a traveling wave in a system with continuously changing mechanical characteristics.

To study how the nervous system supplying the basilar membrane would react to a traveling wave an effort was made to construct a cochlear model with a nerve supply. After some preliminary trials the skin of the lower arm was used to provide the nerve supply. To approach the real situation in the cochlea as closely as possible a large hydrodynamic model was used (Diestel, 1954). As you know, hydrodynamics is that area of physics that contains more paradoxes than all the rest. In this respect it closely approaches the loosest type of psychology that we have today. Yet there is a way of studying hydrodynamical paradoxes by making enlarged or reduced models to reproduce the phenomena.

A model of the cochlea consisting of a plastic tube with a solid wall, and with the interior filled with water, is shown in Fig. 149. On top of the wall was an elastic membrane of varying thickness. This membrane was made stiff near one end representing the stapes, and was soft at the other end of the tube. Figure 149 shows how the thickness of the membrane varied from the "basal" end *a* to the "apical" end *b*, and shows also a rim on which the skin of the arm could be placed. There are many tricks in making such a model operate, which cannot be mentioned here. The complete arrangement is pictured in Fig. 150.

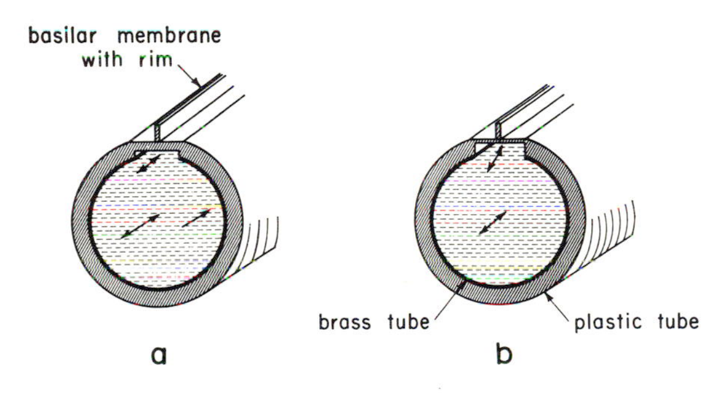

Fig. 149. Cross section of the plastic tube model of the cochlea. From *Journal of the Acoustical Society of America* [74].

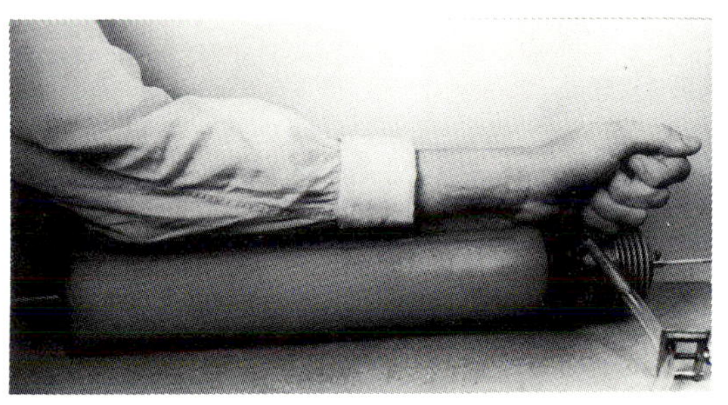

Fig. 150. Plastic tube model with the skin of the arm as the stimulated surface. From *Journal of the Acoustical Society of America* [74].

An interesting feature of this model is that when the piston on the right, where the membrane is stiff, is made to operate the observer feels a vibratory sensation that does not extend along the whole membrane but is located in a narrow section. This section is near the piston ("basal" end) for high driving frequencies and moves toward the other, softer end for low driving frequencies. The amplitudes of vibration of the membrane were measured with a special condenser microphone at the sudden onset of vibrations with a frequency of 125 cps, as shown in Fig. 151. Drawing *a* of this figure shows that the onset was more delayed at points farther along the membrane. This effect represents the time of travel of the traveling wave. Also seen is a maximum of amplitude for the middle portion of the membrane, though the entire membrane vibrates with considerable amplitude. This

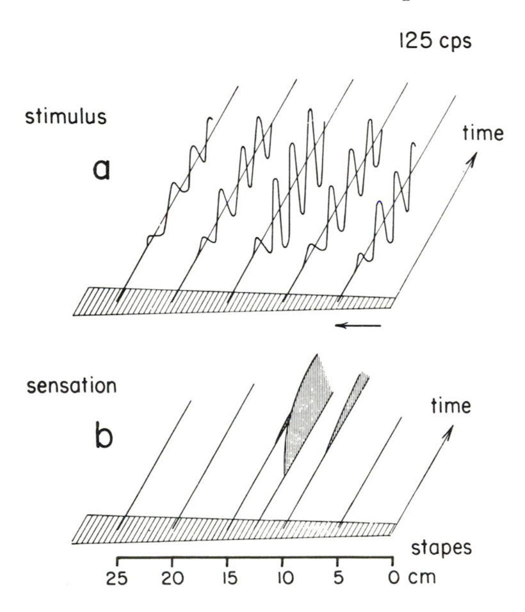

Fig. 151. In *a* is shown the time pattern of vibration of the basilar membrane in the model, and in *b* the resulting pattern of sensation. From *Journal of the Acoustical Society of America* [74].

situation becomes obvious when the vibrations along the membrane are observed under stroboscopic illumination. A traveling wave for vibrations of 125 cps is seen to attain a maximum amplitude near the middle of the tube, and then to decline. The ratio between the maximal amplitude and the smallest one is only about 3:1.

A surprising fact is that when the arm is placed on this model the sensation is distributed along the membrane in the manner indicated in *b* of Fig. 151.

From these observations we must conclude that the mechanical frequency analysis produced in the model and in the cochlea has a very flat pattern. Such a pattern is shown at *a* in Fig. 152, based on measurements made in the human cochlea during stimulation with a tone of 1600 cps.

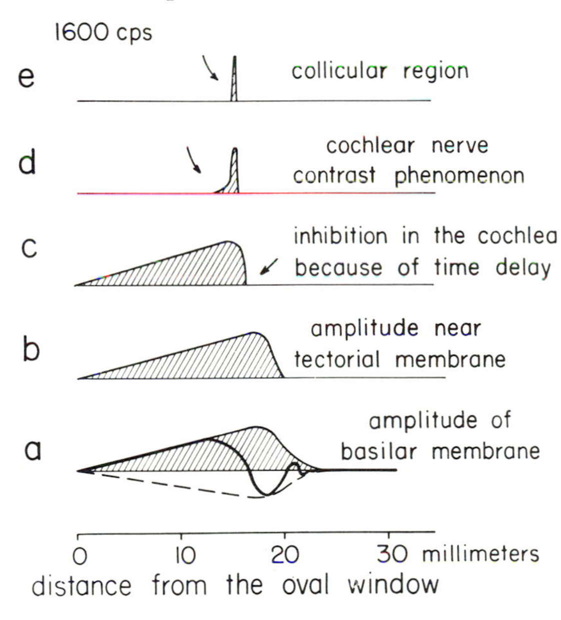

Fig. 152. The presumed sharpening effect of the pattern of response in the cochlea, cochlear nerve, and inferior colliculus.

The amplitudes of vibration of the basilar membrane increase almost linearly with the distance from the stapes up to the maximum and thereafter drop suddenly. If the observation is not of the basilar membrane but of vibrations near the tips of the hair cells near the tectorial membrane, the drop in amplitude close to the helicotrema is even sharper, as shown in *b* of Fig. 152. Hence there is a sharpening effect produced by the mechanical traveling wave that serves to cut off the response in the apical portion of the cochlea.

Because a traveling wave suffers a time delay, the lower frequencies produce a sensation that is delayed relative to a sensation whose stimulus was delivered at the same instant but whose maximum lies closer to the stapes. Under these conditions there is a sharpening of sensation that becomes even more pronounced: to the mechanical cutoff near the helicotrema there is added the inhibitory effect associated with a time delay (see Fig. 152 *c*).

A further funneling effect similar to that responsible for the Mach bands is active on the section of the basilar membrane near the stapes, which further reduces the lateral spread of the localization. This funneling effect seems to continue in the higher levels of the nervous system. In the region of the inferior colliculus the frequency discrimination is quite sharp as a result of the combination of mechanical and neural funneling action.

The most interesting experience that can be produced with the model is the sharp localization of different frequencies of vibration as shown in Fig. 153. This localization is observed

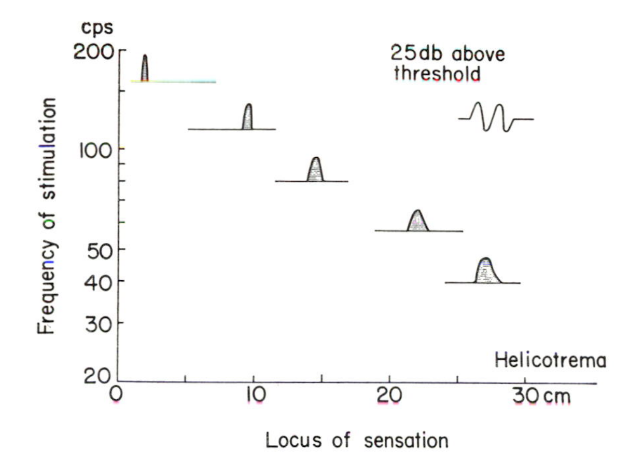

Fig. 153. Variations in the locus of the sensation in the model for different stimulus frequencies. From *Journal of the Acoustical Society of America* [74].

not only for continuous tones but also for stimulation with only two complete cycles of vibration. This is exceptionally rapid localization compared with the frequency analysis that may be made by means of electrical or mechanical filters, which in general have very long onset and decay times. I believe this feature is the best evidence that the cochlear model is a good representation of the actual human cochlea. Savart in 1840 found that the pitch of a tone can be appreciated when only two cycles are used with almost the same accuracy as with a long continuous tone. Savart used a siren to produce these brief tonal stimuli.

The inhibitory action and funneling effect in the auditory system of a cat were demonstrated by Katsuki (1958, 1962) by measuring the threshold of nerve response for various frequencies at different levels of the nervous system. The higher the level the sharper was the threshold response in single neural units as measured with microelectrodes (Fig. 154).

Fig. 154. Single-unit recordings in the auditory tract, *A*, of the guinea pig and at a higher neural level, *B*. After Katsuki (*Journal of Neurophysiology*, 1958, 21, 569-588).

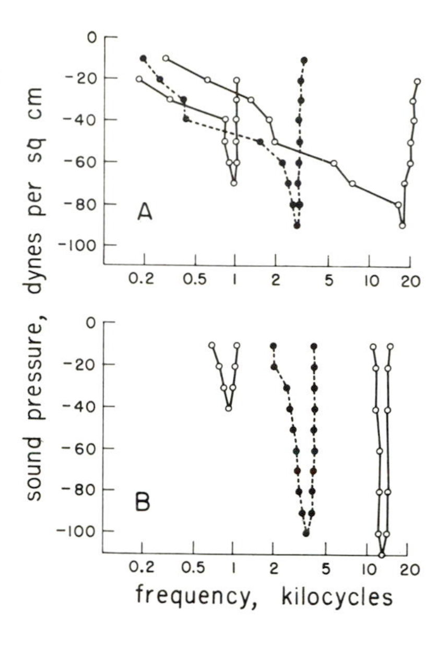

An attempt was made to repeat this sharpening effect on the enlarged model of the cochlea. For this purpose a single nerve unit was imitated by placing only one fingertip, rather than the whole arm, on the vibrating membrane of the model. The observer was asked to determine the thresholds of vibratory sensation over the entire frequency range when the finger was placed on one particular spot of the model. These thresholds were expressed as amplitudes of the driving piston. For the particular spot on the membrane a maximum response was obtained for 80 cps, using the middle finger. This maximum was flat: the responses did not fall off very rapidly as the frequency was increased or decreased. Then when the forefinger was used also, as shown in *A* of Fig. 155, there was a rapid decline for the sensation in the middle

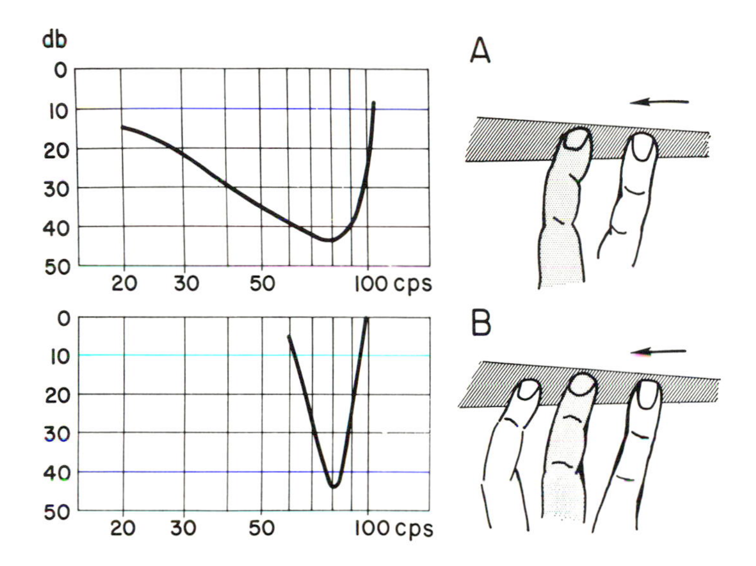

Fig. 155. Sharpening of threshold curve for vibration produced by lateral inhibition. At *A* is shown the curve for the shaded finger when one other finger is near by, and at *B* the narrower curve when two other fingers are stimulated. From *Proceedings 3rd International Congress on Acoustics* [89].

finger as the frequency was raised above 80 cps. There are two reasons for this cutoff in the high frequencies. One reason is that the forefinger receives the traveling wave first. The other reason arises from the fact that the membrane is graduated, and its narrower portions respond more readily to high-frequency vibrations. The forefinger, resting on a narrower region than the middle finger, will be more strongly stimulated at these frequencies. Accordingly, the sensations in the middle finger are inhibited.

For frequencies below 80 cps the inhibitory effect of the first finger is less, until in the extreme low frequencies this effect is only slight. As the frequencies are made lower the time difference in the stimulation of forefinger and middle finger becomes smaller, because the membrane comes more and more to vibrate as a whole, with little phase difference from one region to another. Also for these frequencies such differentiation that remains in the actions of the membrane is in favor of the middle finger, which rests on a wider part of the membrane.

The funneling effect was improved in this model situation by using the ring finger in addition to the other two, as shown in *B* of Fig. 155. Now the amplitude produced by the low frequencies was greater under the ring finger than under the middle finger, so that the ring finger became the locus of the sensation. For this reason, and because in this situation there is inhibition from two sides, the sensitivity of the middle finger is severely reduced, as the figure shows.

There is no doubt that similar interactions take place between adjacent regions of the cochlea during stimulation with a tone, and when two or more tones are present we have to conceive of extremely complex combinations of mechanical and neural activities. In these activities the processes of inhibition and funneling operate so as to produce a frequency analysis of an exceptional kind, in which a train of only two waves at 1000 cps can be discriminated from another train differing in frequency by 0.3 per cent.

The suppression of one's own voice

On first consideration we might suppose that a person is not interested in hearing his own voice. He has control of breathing, the vocal cords, and the position of the mouth, so that he should possess all the necessary information for speaking. What we should really prefer is to hear our own voice in the same way that a listener in the last row of the auditorium does. This matter is important for singers because their success depends largely on the listener's judgment. But unfortunately this reception of one's own voice is impossible. The sound pressure of any generated sound is greatest, in general, at a point close to the source.

When a person speaks there are two ways in which he can hear his own voice. He can hear it by air conduction, because the sounds from the mouth pass around the cheeks to the ear openings, and he can hear it by bone conduction.

Fig. 156. The sound field around the head during the vocalization of three different vowel sounds. From *Journal of the Acoustical Society of America* [53].

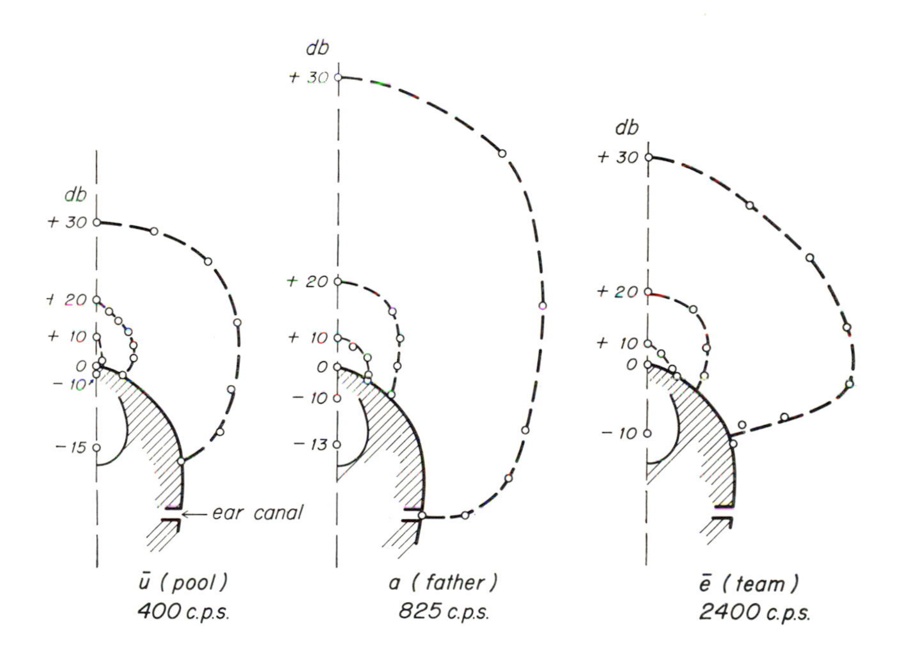

Figure 156 shows results of recording the sound pressures produced by a speaker at points within the mouth and at several points outside.

It will be seen that for nearly all frequencies there is a drop of 30 db or more from the mouth opening to the ear. The broken lines in this figure represent contours of equal sound pressure. From these results we might suppose that the whole important frequency range of speech is transmitted to the external ear opening without distortion and with a reduction of a little over 30 db.

If a speaker wishes to judge the sound intensity of his speech in a room, he will not be interested in the sort of transmission just referred to, but will wish to know about the sounds reflected from walls and ceiling, for these reflections give him a clue as to what the sound intensity in the lecture hall may be. He will need to separate this external sound field from the sound field produced by the acoustic short-circuit between mouth and ear opening. If the reflecting walls do not produce too much frequency distortion, all the frequencies will be reflected with about the same intensity and the reflected sounds will have about the same quality as the airborne sounds passing from his mouth to his ear opening.

Unfortunately, as is well known, a tape recording made of a person's voice always surprises him; his voice sounds very different in reproduction from what he thought it was in speaking. A few decades ago such a recording had a dramatic effect on a speaker, for he discovered that his listeners were appreciating some qualities of his voice of which he was unaware. Usually people were surprised to find that their voices were thin, reedy, and lacking in dynamics.

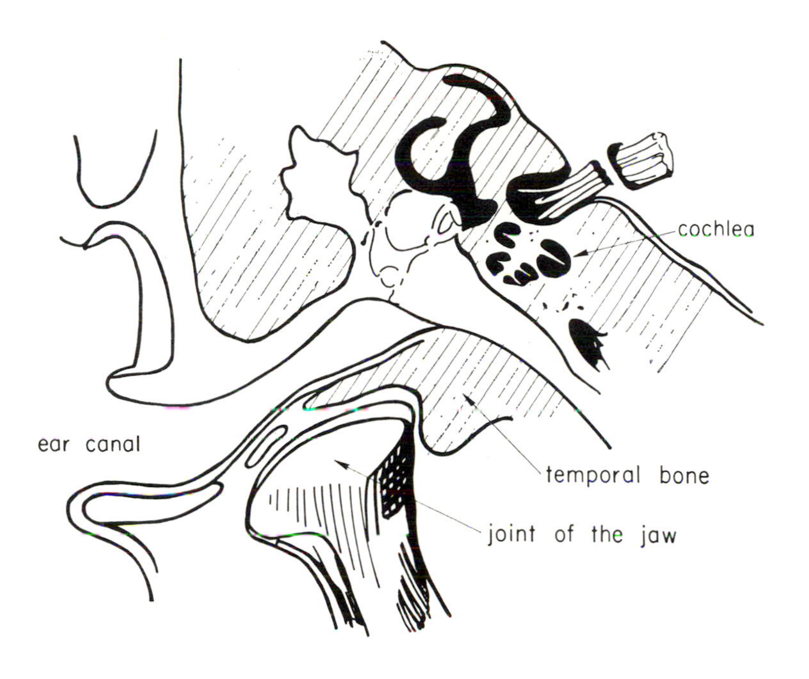

Fig. 157. Position of the mandibular joint near the ear canal. From *Zeitschrift für Hals-Nasen- und Ohrenheilkunde* [40].

The reason for this discrepancy in the hearing of one's own voice is that we hear it by bone conduction as well as by air conduction. As Fig. 157 shows, the joint of the jaw lies against the wall of the ear canal. This ridiculous situation is perhaps a result of our evolution from the fish, because the quadrate bone of the fish became one of the ossicles of the middle ear, and the jaw remained too close to the ear. Any movement or vibration of the jaw is transmitted to the middle ear and to the skull. If we hum a note with the mouth closed and plug the ear opening the heard intensity will increase, as is to be expected from the conditions shown in Fig. 157. Vocalization always produces vibrations of the cheeks, and this vibration is trans-

Fig. 158. Method of measuring the transmission of vibrations through the spine. From *Akustische Zeitschrift* [33].

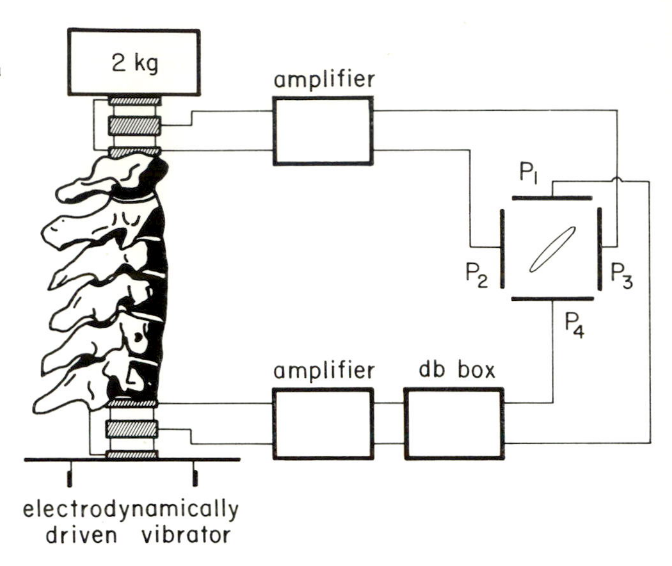

Fig. 159. Pressure transmission across six vertebrae, at various frequencies. From *Akustische Zeitschrift* [33].

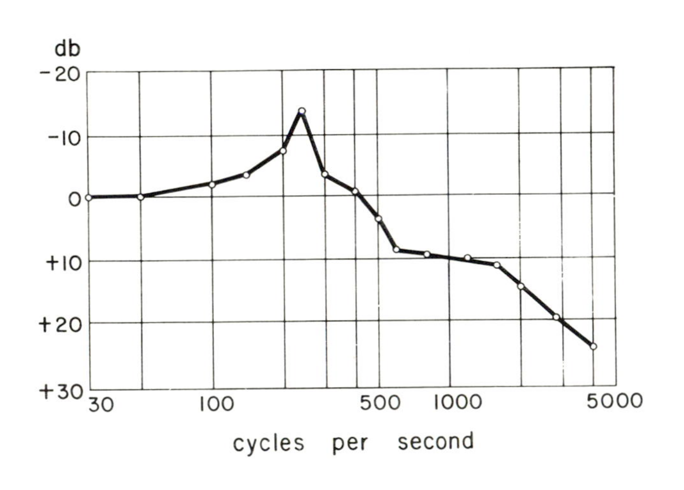

mitted to the jaw. Vibrations are also transmitted through the neck vertebrae to the skull, as may be shown by the method represented in Fig. 158. It is surprising that the neck vertebrae transmit sound pressures within the speech range with relatively slight loss, as Fig. 159 shows. Accordingly, the vibrations of the vocal cords will set the whole skull in vibration, thereby giving hearing by bone conduction.

Hearing by bone conduction is a combined effect of compressions of the entire skull and of the auditory meatus, together with shaking movements of the skull with a complicated frequency pattern.

It is highly frustrating for a trained singer to hear his voice distorted in a strange way by unimportant vocal vibrations in the range below 300 cps. I have read hundreds of pages of instructions on how to sing so that the voice will be localized at a certain place around the head. It is obvious that there is rivalry between two sound fields (as shown in Fig. 160). A singer wishes to suppress the internal sound through inhibition. This is possible because there is a delay of the mouth sounds and of those reflected from a wall relative to the bone-conducted sounds. The inhibition of the bone-conducted sounds can be improved by training, because these sounds are localized closer to the head. When a pure tone is presented to the ear through an earphone on one ear, the sound is localized differently when strong or weak, and when the frequencies are varied. The localization of weak sounds outside the body is represented in the upper part of Fig. 161.

If two earphones are used, one on each ear, and their sounds are made equally loud, the sound is localized in the midline, changing in position according to frequency as indicated in the lower part of Fig. 161. Because hearing of one's own voice is similar to hearing with an earphone, the voice is localized very close to the head. I have read with delight that a good tenor should be able to sing so that he hears his voice not in the throat but on top of the head. This condition holds for the Italian type of singing. The observation may indicate that a good tenor should have a metallic voice with many overtones, in agreement with Fig. 161. Reflections from the wall are localized at a distance, and if the acoustics of the room are properly adjusted for singing, the two sound images will be well separated. For rooms that have too much acoustical damping the reflected sound image is weak and will be masked.

Room acoustics

It was shown in Fig. 145 that a sound wave passing over the heads of a seated audience is greatly absorbed during its propagation because the sound waves are caused to bend downward. This type of wave propagation parallel to the surface of an absorbing material shows some features that are not understood if sound waves are treated only geometrically.

The high degree of damping over the heads of an audience presents one of the problems of room acoustics. A good concert hall must be designed so that not all of the sound waves strike the audience at a narrow angle; other sound waves must reach the audience at a large angle so as not to be damped in this manner. It becomes clear that a concert hall must have two types of sound waves, waves coming di-

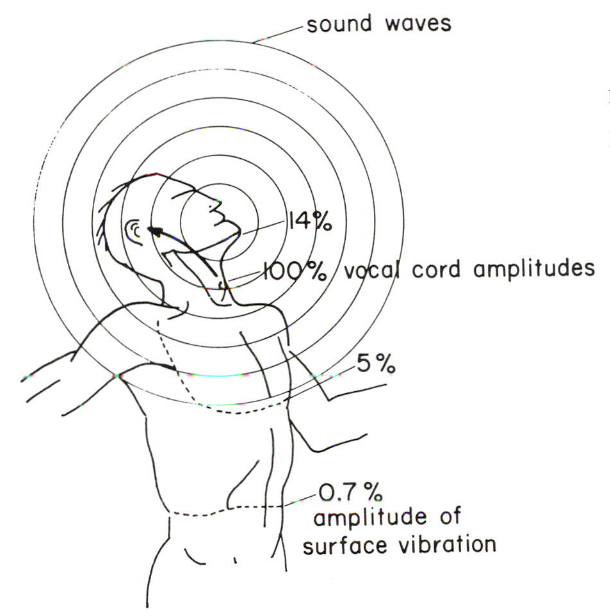

Fig. 160. Propagation of vocal cord vibrations over the surface of the body. The broken lines represent contours of equal amplitude. Revised from *Akustische Zeitschrift* [33].

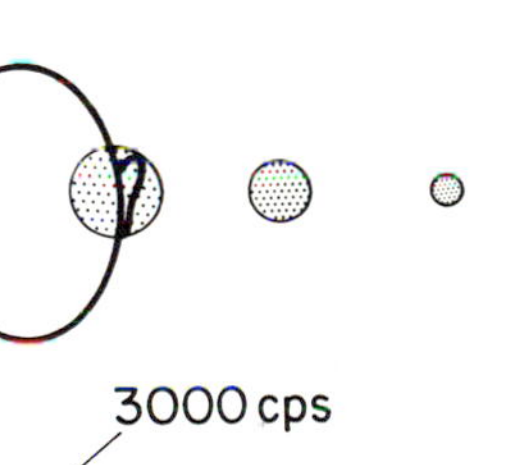

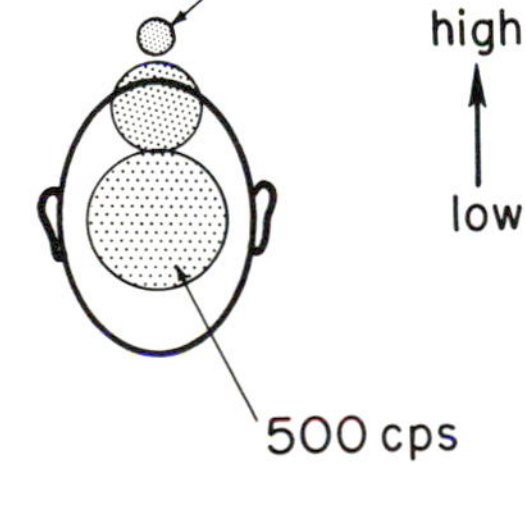

Fig. 161. The localization of pure tones as a function of their intensity and frequency. From *Journal of the Acoustical Society of America* [87].

rectly in a straight line from the sound source, and waves that are reflected either from upper portions of the side walls or from the ceiling, as shown in Fig. 162. This situation indicates the problem involved in the use of reflected sounds for improvement of the acoustics of a room.

The first requirement of a reflected sound is that it must not separate itself from the direct sound. If there is a separation the reflected waves will be heard later in the form of an echo. This rule was formulated largely by Haas (1951). The situation requires that the length of path of the reflected sound cannot be more than a certain amount greater than that of the direct sound. As Fig. 162 indicates, this condition places a limit on the height of the ceiling. In large halls this condition is in conflict with the best visual appearance of the hall.

A second requirement is that the sound energy of the reflected sound must not be too greatly attenuated to be heard. Hence there must not be a straightforward inhibition of the reflected sound by the direct sound on account of their time difference. What is needed is a funneling of the loudness of the reflected sound into that of the direct one. In this type of funneling there will be an enhancement of the direct sound. In addition to the funneling of loudness there is a need of a funneling of the sound direction of the reflected wave into the direction of the direct wave. Otherwise we would localize an orchestra playing in a concert hall as located above the ceiling, for according to the rules of geometric optics a reflection has its image beyond the reflecting surface. This dislocation of the

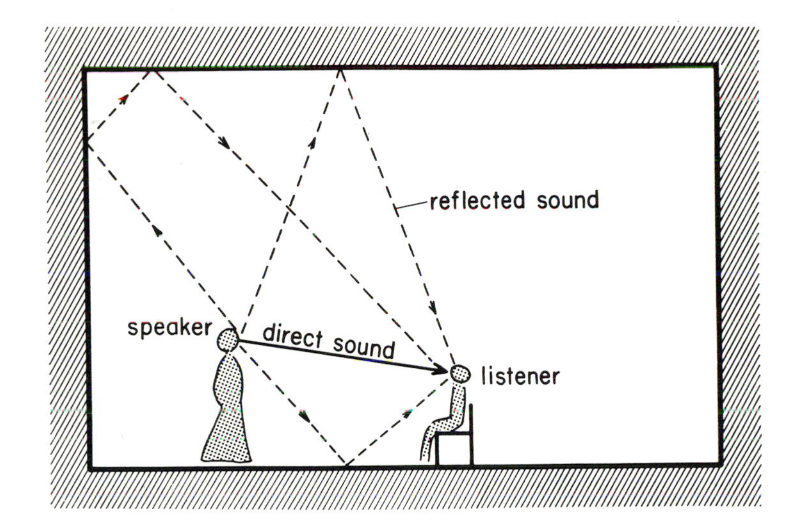

Fig. 162. The paths of sound waves in a room.

sound source may occur for persons in the rear seats of the upper balcony in some halls.

To meet the requirements of funneling it is important that the frequency spectrum of the reflected sound not be too different from that of the direct sound, so that the two sounds may readily mix and combine their loudnesses. Funneling of the reflected sound into the direction of the direct sound may be difficult to achieve if there is only one reflected sound, but becomes much easier if there are many reflected sounds coming from all directions. (Such sounds are called reverberations.) The direct sound, because of its temporal priority, can readily determine the localization of the sound source. The problem of room acoustics becomes critical when economy demands that the hall contain as many seats as possible. This is an ancient problem, which may be observed even in the construction of Greek amphitheatres.

There is a certain periodic pattern of time in the activity of consciousness. One of these periodicities is 0.8 to 1.2 seconds long, and it interjects a rhythm into the observations and recognitions of music, singing, and speech. We tend to break both speech and music into temporal segments of this duration. This temporal pattern operates so as to enhance the reflected sounds that arise within one time unit and to suppress the sounds that come after the unit has ended.

Another phenomenon of importance in room acoustics has to do with the angle from which the sounds come. We are unable to listen with the same degree of concentration to sounds coming from the front and to other sounds coming at the same time from behind. Usually we switch our attention periodically from front to back. Closer observations show that the spatial angle within which we are able to integrate our perceptions is not the whole front or back region but is limited to a cone of about 30° to left or right or up and down. If a sound source is suddenly presented in a strongly reverberant room, as in a church, these spatial angles are readily noticed.

First is heard a loud sound with a very narrow angle about the midline of the head, and then a widespread reverberating sound is heard about a second later coming from the walls. Another second later we may hear the reverberation from the dome of the church. This observation shows that we are able to inhibit sounds coming from directions that are unimportant to us, and to shift our attention from the small central cone to larger angles in the room. A similar effect is found at a party when we are surrounded by a large number of people talking with about the same intensity. We have no difficulty in concentrating our attention on a speaker in front of us and in inhibiting the talk of other persons to unrecognizable noise.

An experiment was carried out to measure the spatial angle within which sounds are combined to produce a common sensation. The sounds outside this angle cannot be integrated, and these (or the others) must be inhibited. It has turned out that most of the phenomena of room acoustics can be imitated by vibrations on the skin, and a useful arrangement for this study is shown in Fig. 163. On the left is a noise generator operating into a Wheatstone bridge. The attenuators shown were set so that when the switch was open the energy of the

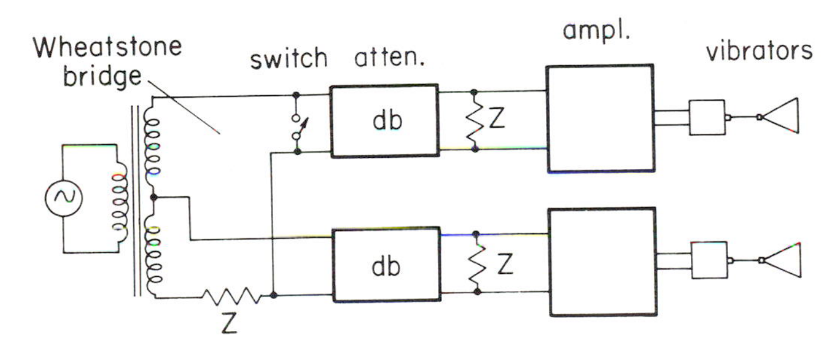

Fig. 163. Bridge arrangement for switching on two vibrators alternately without intermission. From *Journal of the Acoustical Society of America* [83].

generator went into the upper channel and when it was closed the energy went into the lower channel, and without any time delays. This condition was obtained by proper adjustments of the input impedances of the attenuators. With this circuit it was possible to switch instantly, without interruption time, from one vibrator to the other.

If the two vibrators were made to act on the skin surface as shown in *a* of Fig. 164, and were located close to one another, it was possible to switch from one vibrator to the other every 1/4 second and still have the impression that only one vibrator was present producing a continuous sensation. The distance between the vibrators was such that the two vibratory sensations were integrated. But an increase in the distance between the vibrators, as shown in *b* of this figure, made it no longer possible to fuse the two sensations into one, and only one or the other was concentrated on. A pulsating sensation was felt either at one vibrator or the other, but never was there a continuous sensation. This observation indicates that under certain conditions the field of perception for the skin is limited, and sensations outside of this field are inhibited.

The same phenomenon can be produced with this equipment if the vibrators are replaced with loudspeakers placed about 10 feet in front of an observer. When the speakers are side by side the observer hears a continuous sound coming from a point about midway between them. But if the speakers are displaced laterally a distance will be reached at which a pulsating sound will be heard coming from either one speaker or the other. Again the evidence shows that there is a limit to the angle of our summation of sounds in space.

As concert halls are enlarged it becomes increasingly difficult to funnel the reflected sounds into the direction of the direct sound. One rea-

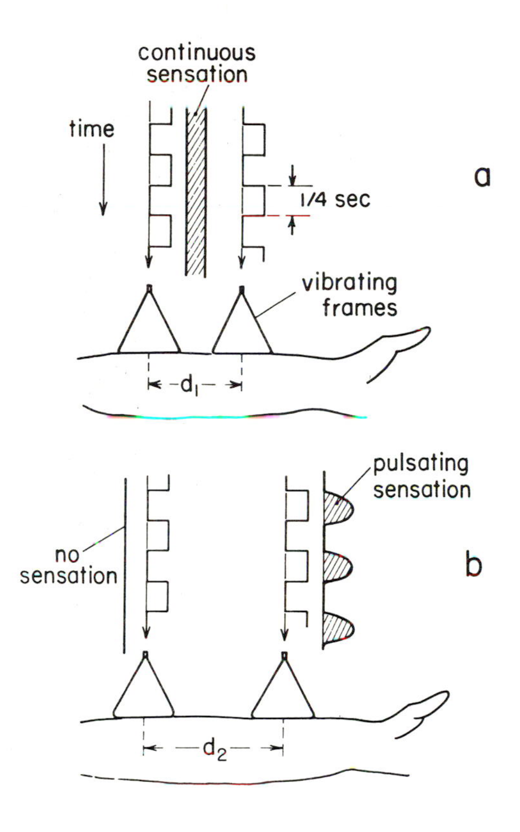

Fig. 164. The effect of distance of separation on two alternately applied vibrations. From *Naturwissenschaftliche Rundschau* [113].

son is that the reflected sounds may have a different quality. The qualitative changes are caused by the presence of sound-absorbing material placed on walls and ceiling, because porous materials absorb high frequencies to a greater extent than low frequencies. This difference in quality is further increased by the fact that the loudness of the reflected sounds is different from that of the direct sounds. Because loudness changes are different for low and high frequencies for the same change in sound pressure, there is a frequency distortion. These differences are due to the peculiarities of the ear itself.

The importance of this loudness change can be demonstrated by comparing the common use of absorbing material with the reduction of sound intensity that may be obtained by use of an open window, which conducts a part of the sound to the outside without any frequency selection. Also a comparison may be made by the use of nonporous materials, which do not have the same selective properties as the materials commonly used. I carried out an experiment of this kind on a large scale by placing an orchestra in a court as shown in Fig. 165.

In this situation there were no reflections from above or from a portion of one side. Nearly all observers, and all the musicians, agreed that the musical quality was much better than in the concert hall where this orchestra usually played.

The experience was similar when a large tent was used. Here also the absorption was similar for medium and low frequencies, and it was astonishing to observe the clarity of the understanding of spoken words and the trilling of some of the musical instruments, even at a great distance.

From this evidence the conclusion was drawn that two variables must be adjusted simultaneously to bring a concert hall to maximum effectiveness. It is necessary to adjust the ratio of direct to reflected sound, and to take account of the frequency dependence of sound absorption on walls and ceiling.

In carrying out these experiments, two types of sound-absorbing material were used. One con-

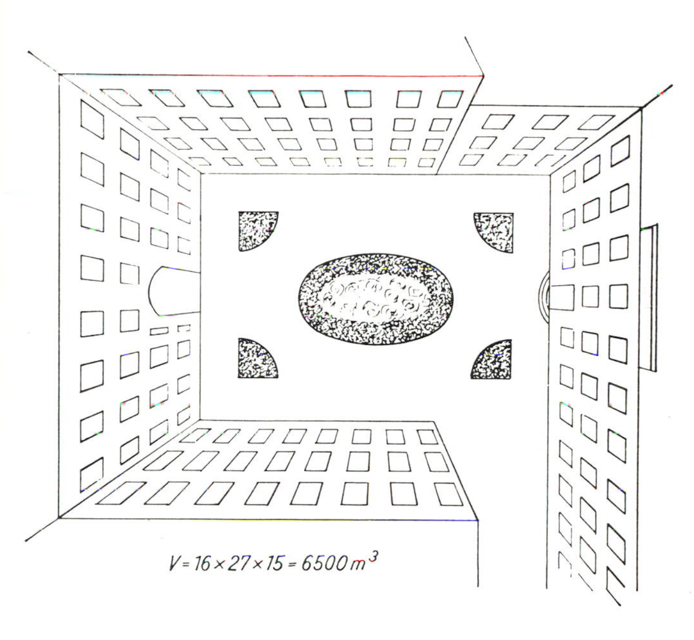

Fig. 165. A courtyard with superior acoustics.

sisted of the usual porous materials, as shown in *a* of Fig. 166, which absorb mainly the higher frequencies. The other was a membrane type of material, shown in *b* of Fig. 166, which mainly absorbed the low frequencies and reflected the high frequencies without loss. This action is similar to that of the canvas of a tent.

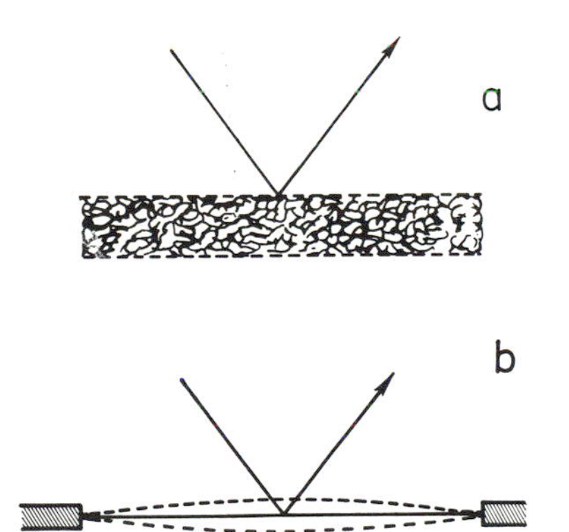

Fig. 166. Two types of acoustic absorbing materials: *a*, a cotton or glass fiber batt; *b*, a plastic membrane.

Many of the experiments were carried out in a room, shown in the photograph of Fig. 167, with a volume of 2000 cubic meters. The orchestra was at the back of the room and the music experts, some of them well-known conductors, were seated at the front. A long list of qualities was made up and marked by the observers according to their degree of agreeableness, for various acoustic treatments of the room. A group of helpers moved the absorbing material on the side walls in and out of the room. All the experiments used the method of limits, reaching an optimal adjustment by going alternately from the least amount of absorption to the largest amount. Thus we placed in the

Fig. 167. Procedure for adjusting a concert room for optimum acoustics.

room an excess of a cotton type of absorbing material until it became obvious that the high frequencies were lost, and then exchanged this material for the low-frequency absorbing type until the low frequencies were lost. This procedure was carried back and forth to obtain a condition that was considered free of distortion.

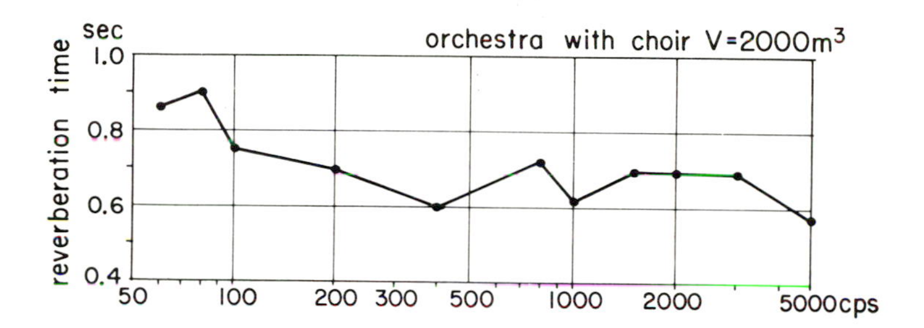

Fig. 168. The optimum reverberation time for various frequencies as obtained in the room shown in the preceding figure.

The reverberation time for this room as a function of frequency is shown in Fig. 168. By reverberation time is meant the time necessary for the sound pressure of the reverberant sound field to decline by 60 db after the sound source was switched off. The reverberation time is a measure of the effectiveness with which sound waves were absorbed on the walls and ceiling.

A room of 2000 cubic meters is a small one, and there is no difficulty in adjusting the shape and absorption of such a room to have satisfactory acoustical properties for a single singer, a quartet, a salon orchestra, or a small orchestra. But it is quite obvious that for every room there is an optimal number of musicians or singers. Even in this small room it was possible to have situations in which the funneling of the reflected sound into the direct sound was no longer effective, and the sounds began to be perceived as coming from a place above the ceiling. The best way of avoiding this difficulty is to introduce panels that break up the surface, such as may be seen in Fig. 167.

In large rooms this difficulty increases, and there is always a tendency to place special reflectors around the sound source or at one side of it so that the room will seem visually large but acoustically smaller. Reflectors located below the ceiling have been tried for this purpose also. Many experiments of this kind have been carried out recently. Some have been successful at times, but not very satisfactory under all conditions. The reason is that the room acoustics are being corrected for the inhibitory and funneling properties of the ear, and these phenomena can sometimes play tricks.

When I was a beginning telephone engineer, we had at hand some long-distance cables with practically no distortion in the range between 300 and 3300 cps. We supposed that because there was no frequency distortion we could switch more and more cables into service and have telephonic communication at extremely great distances. This plan did not work at all. The physical measurements showed that there was no frequency distortion for the long cables, and because this was the main way of testing cables, all seemed well. Yet speech was extremely difficult to understand over such cables because at the beginning of every vowel the low frequencies masked the high frequencies in a manner that was not expected on a basis of the electrical qualities of the system. The reason for the masking was disclosed when we suddenly switched on an alternating current of 1000 cps, as seen in Fig. 169. The onset of the transmitted signal began with very low frequencies, and only

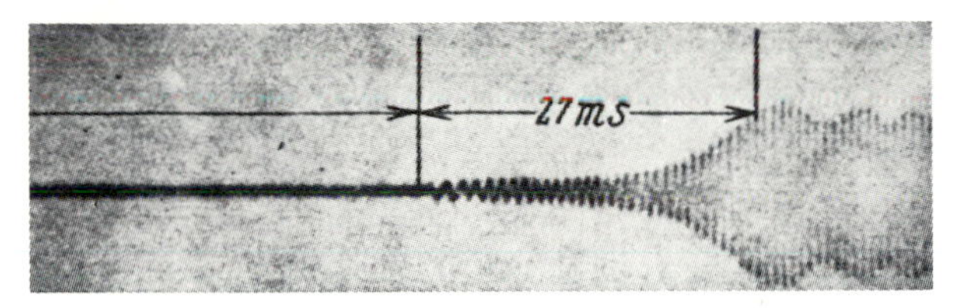

Fig. 169. Phase distortion in a long-distance cable of an early type. From Küpfmüller (*Die Systemtheorie der elektrischen Nachrichtenübertragung*, 1949; S. Hirzel Verlag).

after a time did the 1000-cps tone come through (Küpfmüller, 1949). This behavior could easily be understood, for it had long been known that in this type of transmission line the low frequencies travel at a greater speed than the high frequencies, and the difference could be very striking. As Fig. 169 shows, there was about 25 to 30 milliseconds of delay between the full onset of the low-frequency and of the 1000-cycle tone. Therefore every vowel began at a low pitch and then went over partly to a higher pitch. The real difficulty, however, was that the low-frequency components that arrived first could inhibit the higher components. For a continuous sound the components arrived in proper relation, but for short pulses of tone the lower frequencies were more prominent because of their early arrival.

This defect is called "phase distortion," and though it changed completely the design of long-distance cables it was not well recognized in the field of hearing. I was surprised when using condenser loudspeakers with large surfaces at the excellent musical quality in comparison with electrodynamic loudspeakers. The condenser speaker has practically no transients in the frequency range of interest, whereas electrodynamic ones have large transients. Also condenser loudspeakers have no phase distortion, whereas electrodynamic speakers do, and the difference is clearly audible.

To improve the acoustics of a large concert hall in New York City a number of reflecting panels were installed just below the ceiling, as shown in Fig. 170 (Beranek, 1962). Such panels when placed in the proper position reflect the high frequencies to the places in the audience where there is a need for increases of sound level. This method has been regarded as offering the possibility of increasing the size of concert halls.

For ordinary wave propagation it is known that an obstacle such as a panel will produce a reflection of the short waves according to the usual laws of geometric optics, but low tones whose wavelengths approach the dimensions of the obstacles will be reflected only slightly, and instead will bend around the obstacle and continue in their original directions. This behavior

Fig. 170. Reflecting panels on a ceiling. After Beranek (*Music, Acoustics and Architecture*, 1962; John Wiley and Sons).

is shown by waves on the surface of water, as seen in Fig. 171 (Pohl, 1932). It is a well-known phenomenon and explains in part why the sky is blue and the sun is reddish. The short wavelengths of light from the sun are scattered randomly by the various particles in the air. But the long waves, which are seen as red, are not scattered to the same extent and follow their original path, forming a red image of the sun.

Fig. 171. Wave patterns produced by an obstacle. *a* shows an unobstructed sound field; *b*, the shadow cast by a small obstacle; *c*, the denser shadow of a larger obstacle; *d*, the reflections produced by an obstacle at an angle. After Pohl (*Mechanik und Akustik*; *Einführung in die Physik*, 1932, Springer-Verlag).

As may be seen from the cross section of a concert hall shown in Fig. 172, the reflecting panels will convey high-frequency components of sound to a listener by a shorter path than that taken by the low-frequency components. The low-frequency waves will bend around the panels and pass on to the ceiling, to be reflected there. Hence the high frequencies will reach the listener first, and the low frequencies a little later. The situation is the reverse of that encountered in the long-distance cables, in which the low frequencies arrived first (Fig. 169). Yet in both instances there is inhibition for much

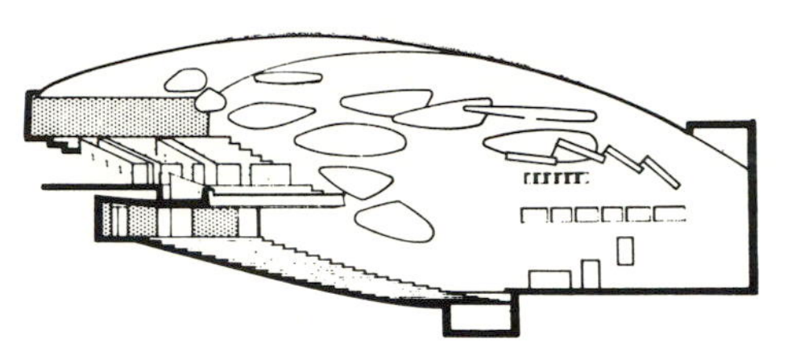

Fig. 172. Cross section of a concert hall. After Beranek, (*Music, Acoustics and Architecture*, 1962; John Wiley and Sons).

the same reasons. In the concert hall the low frequencies will lose their effectiveness in brief pulses of sound as heard by a listener. These tones are physically present but are inhibited because they are delayed in arrival at the listener's ears. This type of distortion may be called "room acoustics phase distortion" and represents an interesting new field. The construction of auditoriums and the subsequent discovery of these distortions represent very expensive experiments on inhibition.

The failures in long-distance cable telephony were just as expensive as those of room acoustics, but have not been made known to the public. Such failures must be expected if we are optimistic and wish to make progress. Phase distortion is something that most people do not recognize and are not annoyed by as long as it is slight. Therefore when, under economic pressure, the scale of construction is increased, this distortion is also increased and suddenly the unexpected happens: we find that there is a threshold for phase distortion just as for other sensory phenomena. If we exceed the threshold most people will be aware of the fact and if the effect is unpleasant they will complain. The sensory thresholds present barriers that often prevent technical extensions into what otherwise might be unlimited fields.

VI · THE ROLE OF INHIBITION IN VARIOUS FIELDS OF SENSORY PERCEPTION

Temporal quanta and the periodic inhibition of sensations

In this final lecture I should like to discuss the many aspects of inhibition in relation to some of the old familiar problems of psychology and behavior. Now that we are aware of the variety of manifestations of inhibition it is possible that its applications to these problems will provide new points of view for their understanding. This approach has heretofore been handicapped by the assumption that the inhibitory and funneling properties of the nervous system produced only minor side effects. The direct input-output formulation was regarded as representing the main phenomenon. However, as our technical capabilities improve we are brought closer to the view that without inhibition there would be no way of handling the great quantity of information that we receive. This situation obtains for man and other animals alike. I should therefore like to present a few problems in which inhibitory processes play a role that has so far been little explored. Many of these problems are for the most part of central origin.

It is well known that we are not able to observe continuously or even to think continuously. We do these things in certain intervals of time that may be called "temporal quanta" (Stern,

1897; Lehmann, 1905) or the "conscious present."

This phenomenon can easily be observed by presenting to an observer a pure 1000-cycle tone about 40 db above threshold, maintaining it continuously at this level, and asking the observer whether it exhibits any changes. Most observers say that they are unable to concentrate steadily on the loudness of the tone, but can note it only for certain periods of time. Between these successive periods there seems to be a certain momentary lack of consciousness when the loudness of the tone decreases. The length of the period is about 0.8-1.2 seconds. The length is not stable and sometimes the lapses of consciousness are hardly noticeable, whereas at other times they are quite obvious. In general, however, this phenomenon divides the whole temporal scale into small sections. From the point of view of electrophysiology, I have always been interested in these momentary lapses of consciousness in which momentary reductions in the magnitude of sensation occur. Yet in the records of neural activity, for single nerve units or larger groups, there does not seem to be any counterpart of this effect. We find for a continuous stimulus an uninterrupted series of nerve discharges. The momentary lapses seem to be of central origin, and they can be regarded as momentary, periodic inhibitions of central activity. Because there is a certain degree of correlation between this periodicity and the heartbeat, it is possible that a lapse arises from a momentary reduction in the blood flow to the brain. Often, however, there is no relation of the periodicity to the heart rate, and the possibility arises that when a relation exists it is due to an action of the periodicity of the heartbeat on the central nervous system or the reverse. There is the possibility also that the electrical activity in the brain is not continuous, that there is, rather, a periodic scanning of the entire cortex. The presence of temporal quanta was suggested a long time ago, and there are many ways of observing them.

One of the earliest experiments was to beat out with the finger a rhythm regarded as neither fast nor slow. It is interesting to note that the rhythm then produced is usually well defined, and may be maintained for a long while even when some disturbances are introduced. Figure 173 shows the intervals between successive taps produced by two different observers over a period of 15 minutes.

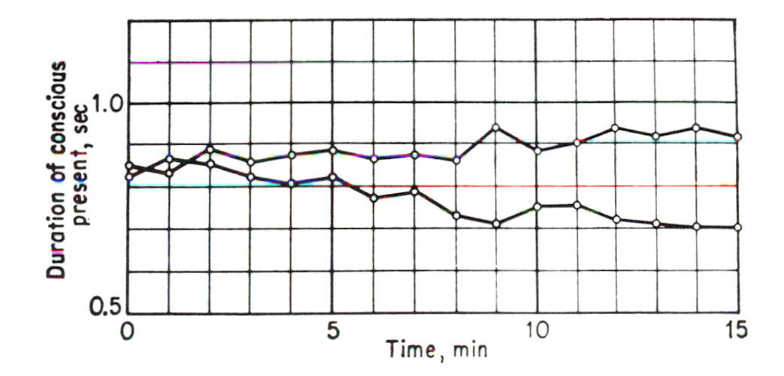

Fig. 173. Rates of tapping by two persons instructed to produce rates neither too fast nor too slow. From *Annalen der Physik* [9].

Another way of observing this form of central inhibition is to record the number of syllables per word produced during continuous speech. Vierordt in 1868 noted that there is a definite preference for words of two syllables; about 45 per cent of all words used are disyllabic, as shown in Fig. 174. The time required to pronounce a two-syllable word corresponds roughly

Fig. 174. Syllabic frequencies for verbs and nouns.

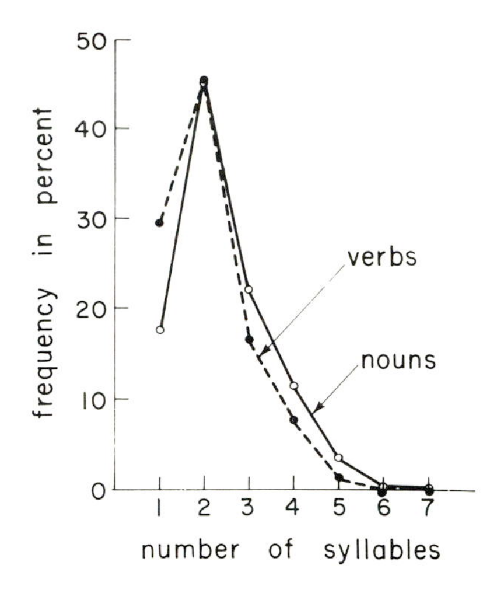

to the temporal quantum defined above. Syllables pronounced during this period are likely to fuse into a single unit. In speech there is a definite tendency to break the discourse into certain chunks and to try to recognize the chunks as units, disregarding the fine structure within the chunks. This fact may be shown experimentally by presenting a series of clicks to an observer and then slowly beginning to increase the click rate by an automatic method. The observer first will combine two clicks, and then three or four into particular groups, and will listen to these as a group without being conscious of the single clicks as such. In hearing, and especially in speech and music, this separation of a continuous process into small, discrete periods of time seems to be extremely important.

In a study of the decay time for sensations of loudness, most observers were found to have no difficulty in following continuously the decay of

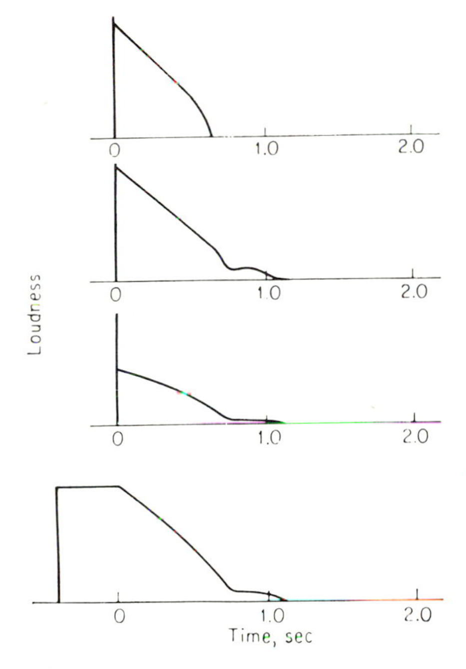

Fig. 175. Effect of the conscious present on the decay of loudness of a tone. The first three curves are for different observers. The lowermost curve indicates that when the tone was sustained for a time before its decay began the perceived effect was as before; the interruption in the smooth course of loudness decay caused by the conscious present occurred after about 0.8 seconds as in the other instances in which the tone was not sustained. Redrawn from *Annalen der Physik* [9].

loudness, and they had the impression that the decline in loudness was greater at the end of each temporal quantum. This situation is represented schematically in Fig. 175. If the decay time exceeded 1 second then in general there was a sharp breakdown of the magnitude of the sensation at the end of a quantum when consciousness showed its momentary lapse.

Perhaps an even more interesting fact is that we are willing to make a fresh start after each lapse of consciousness. It might be said that we store the information obtained during the preceding quantum and concentrate on the next one. When phenomena are complex this activity produces a periodic shift of interest from one sensation to another. The shift of interest may be also a shift in conscious space, so that we begin to make a new localization of the sensations after each lapse. It is easy to observe this phenomenon when a large series of sound sources is present and we attempt to observe them all. In general we shift from one sound source to another in a rhythm corresponding to the temporal quanta.

Central inhibition was one of the first forms of inhibition to be discovered, yet little progress has been made in the psychological observation of it. In hearing especially there are examples of central inhibition that have been known for centuries. A familiar example is the miller who awakes when the mill stops. He is able to inhibit all the normal noises made by the mill and his sleep is not disturbed. He wakes up when the routine inhibitions stop because the neural processes produced by the usual stimulation are no longer present.

Projection of a sensation outside the body

The funneling of sensations into a space outside the body is an important feature of neural funneling, for it controls practically all our behavior. For example, reflected light from an external object produces an image on the retina. The sensations exist only within our body, yet we localize the image outside the eye, even when we use only a single eye and look at an object far away. This localization beyond our perceptual system is of great importance for survival because it enables us to appreciate impending danger or objects of great necessity. This externalization is achieved without the slightest recognition of the optic image itself or the stimulations on the retina.

The same conditions hold for hearing. The sensations are produced by the action of stimuli on the basilar membrane of the cochlea. The cochlea is deeply imbedded in bone, but we do not localize auditory sensations there but usually refer them to a source somewhere in the environment. However, as we have seen, this external reference does not seem to be true for hearing with earphones.

This external projection has probably been learned early in life; certainly this is true for hearing and vision. But we have not acquired this kind of external projection for skin sensations, and so we have an opportunity to discover how stimulus projection in space is learned.

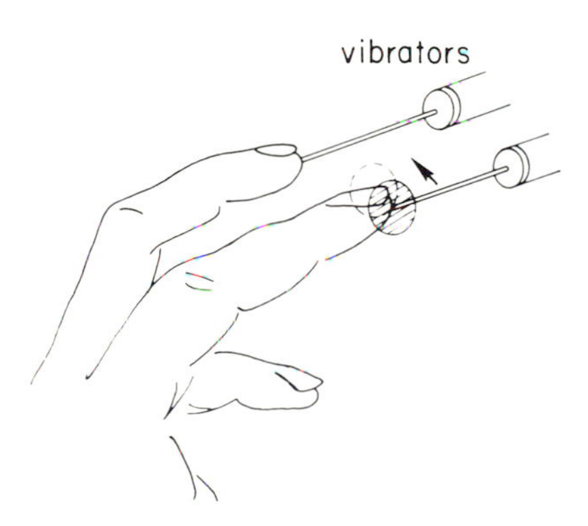

Fig. 176. The localization of a vibratory image in the space between two fingertips.

For this study a pair of vibrators stimulate two fingertips as in Fig. 176. Each vibrator is actuated by the same series of clicks, and their applied currents are varied to give equal magnitudes of sensation on each fingertip when the stimuli are presented separately. Also the setup includes a means of varying the delay time between the clicks of the two series. If a click is delayed for one finger more than 3 or 4 milliseconds, a person feels separate sensations in the two fingertips, as already described. If, however, the time between clicks is reduced to about 1 millisecond the two click series will fuse into one, and the vibratory sensation will be localized in the finger that receives each click the earlier. If the time delay is further decreased the sensation for a trained observer will move into the region between the two fingers, and if then the time relation between the two click series is reversed the click will move to the opposite side. For a person who observes such stimuli for the first time, the click will make a sudden jump from one fingertip to the other on the very first trial. The movement is not continuous from one finger to the other. The jump is abrupt and the observer can localize vibratory sensations only beneath the vibrating tip.

After a few days of observing vibratory phenomena, however, the observer finds that the shifting of sensations on the fingertip, and elsewhere on the skin, loses its abrupt character and goes over into a sort of creeping from one vibrator to the other. Usually after two or three weeks of training the observer can experience a continuous motion of the vibratory sensation from one finger to the other, with a rate of motion corresponding almost linearly to the time difference between the two clicks. This new experience develops more readily when care has been taken that the two vibrators produce stimuli that are identical in magnitude.

The interesting point in this experiment is that for the condition in which there is no time delay the vibrations are localized between the two fingers where no skin is present. If the fingers are spread apart the same effect is found, and when the amount of time delay is varied the sensation will move correspondingly in the free space between the fingers.

Even more dramatic than this experiment is the one in which two vibrators are placed on the thighs, one above each knee. Here the vibrators can stimulate large skin surfaces and produce strong vibratory sensations. By training an observer first to note the localization of the vibration when the knees are together, he can be made to perceive a sensation that moves continuously from one knee to the other. If the observer now spreads the knees apart he will again experience at first a jumping of the sensation from one knee to the other. In time, however, the observer will become convinced that the vibratory sensation can be localized in the free space between the knees, and he will be able to experience a displacement of the sensation in this free space when an appropriate time delay between one stimulus and the other is introduced. This experience is a very peculiar one.

We can go even further in this projection of sensations into free space with the equipment shown in Fig. 177. In this setup are two microphones connected to vibrators that rest on the right and left sides of the chest. When a loudspeaker producing clicks at a rate of about 2 per second is placed on the left as shown, a vibratory sensation will be felt on the left side of the chest. Then if the loudspeaker is moved from left to right the vibratory sensation will

Fig. 177. Equipment for study of the localization of vibration outside the body. Sound from a loudspeaker is picked up by two microphones and converted into vibrations. From *Annals of Otology, Rhinology, and Laryngology* [102].

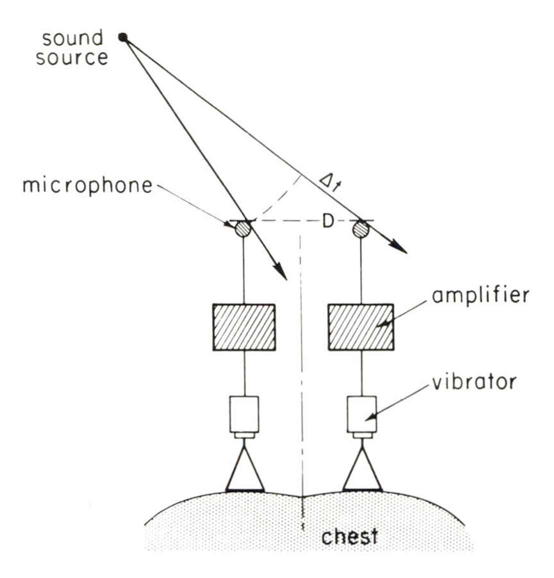

jump to the right side of the chest. Whenever the sound source is on the extreme right, we feel the vibrations only on the right side of the chest, and when it is on the extreme left we feel them only on the left side of the chest.

If the vibrators are carefully matched, this jumping from one side of the chest to the other occurs precisely in accordance with the position of the sound source. After weeks of practice an observer will no longer experience the jumping of the vibratory sensation from one side to the other, but this sensation will move continuously in accordance with the lateral position of the sound source. He can look at the sound source and compare it with the location of the vibratory sensation to prove the close relationship between them. However, even after days of experience of this kind, the vibratory sensation will continue to move and be felt on the surface of the chest.

Some observers, however, following with the eye the position of the loudspeaker in the room and the close agreement between their angle of view and this position, became able for no obvious reason to locate the vibratory sensation close to the loudspeaker when its distance from the chest was no greater than 2 or 3 feet. It is difficult to say how this distant projection of the vibratory sensation was achieved. In general there was an intermediate stage at which both possibilities were present, when the vibrations could be localized on the skin surface or be projected to the position of the loudspeaker. There is no doubt that the visual observation of the loudspeaker position played a significant role in the learning process. After several months of training it became possible to localize the vibratory sensation, even with the eyes closed, at a position outside the body, though usually this position was closer to the body than to the loudspeaker. No observer believed that he would be able to project the vibratory sensations farther than 3 feet from his skin, even after further training.

This matter of the external projection of vibratory sensations seems to be strange and hard to believe, yet it is well known in many fields. Every well-trained machinist projects his sensations of pressure to the tip of a screwdriver, and it is this projection that enables him to work rapidly and correctly. For most people this projection is so common that they are unaware of its existence. The same type of projection occurs in cutting with a knife, and our adjustments of the blade make use of sensations projected to its edge. Nowadays the palpation of tumors and cysts does not play the same role in medical diagnosis that it did before the extensive use of X-rays, but in earlier times a good practitioner did not feel a tumor as at his fingertips but he projected his vibratory and pressure sensations into the patient. His procedure used a complex interaction between pressure and vibratory sensations in locating a small tumor cyst, and the location could later be verified in surgery.

I found the localization of sensations in free space to be a very important feature of behavior. To study the matter further I wore two hearing aids that were properly damped so that the sounds could be picked up by means of two microphones on the chest and then transmitted to the two ears without change in pressure amplitude. Stereophonic hearing was well established, but a perception of the distance of sound sources was lost. I shall not forget my frustration in trying to cross the street during rush hour traffic while wearing this transmission system. Almost all the cars seemed to jump suddenly into consciousness, and I was unable to put them in order according to their immediacy. I should probably have required weeks of experience to become adjusted to this new type of projection. A small change in the amplification of one side was enough to cancel the whole learned adjustment.

From these observations I reached the conclusion that the localization of sounds by persons suffering from disorders of hearing is just as important as their hearing of the sounds. If we listen to sounds with only one ear, we usually localize them close to the head. This is especially true for pure tones.

There is no question that localization plays an important role in our perception of internal sounds produced in our own body as distinct from external sounds that are vital for survival. The more we are trained to project external sounds outside the body the easier it becomes to inhibit internal body noise.

Certain fishes are known to detect foreign objects by producing an electric field in the water and observing its changes on their approach to the object. It is surprising how precise this localization is. Lissmann and Machin in 1958 investigated this problem. Figure 178 shows the changes that take place in the electrical field for foreign bodies with different electrical resistances. The fish has an electric field around his body that is modified by his body's presence. A change in the field caused by the approach of a foreign body has to be distinguished from the pattern of his own field. It is likely that in this instance the distortions of the field are projected outside.

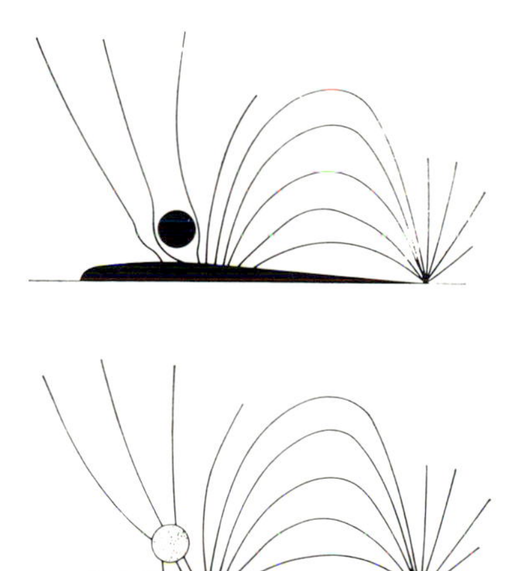

Fig. 178. Patterns of current flow between head and tail of an electric fish, used in the localization of objects with different conductivities. After Lissmann and Machin (*Journal of Experimental Biology*, 1958, 35, 451-486).

Many fishes are able to recognize foreign objects through a distortion of the flow of fluid around their bodies. They can easily identify the approach of the glass wall of an aquarium without bumping into it even when vision is absent. At present I can only admire these achievements, for I have no explanation of them. The most fascinating animal is the porpoise, who leaps after some object held high in the air. The animal must change his visual projection from one appropriate for underwater vision to another for air vision.

Optical illusions and lateral inhibition

Earlier it was shown in the mathematical treatment of the Mach bands that almost the same forms of these bands were produced when the inhibitory unit procedure was applied not just once but three times consecutively. The distribution of the sensory pattern obtained by this repeated manipulation comes so close to the one-step operation that we have no way of telling whether in vision the Mach bands are produced in one step or in several. It is highly probable that several repeated steps are used. This consideration brings up the question whether some of the optical illusions, which are modifications of a stimulus pattern, may not be understood by considering them from the point of view of lateral inhibition. As Fig. 68 shows, in vision the neural unit has a wide inhibitory area, so that the lateral inhibitory effects probably extend to large regions. It is possible, however, that the lateral spread of the neural unit is different at different neural levels.

It was shown by Révész (1934) that optical illusions have their counterparts in skin sensations. Several of his experiments were repeated using cardboard cutouts of various shapes pressed on the skin surface, and comparing the pressure pattern thereby produced with the resulting pattern of sensation on the skin surface. A major difficulty in working with illusions on the skin is that the difference limen for distance discrimination is large, amounting to several centimeters, and often there is not enough room on the skin to produce conveniently the stimulus patterns of a size comparable with those used on the retina in optical illusions. Nevertheless it is possible to produce some illusions on the skin surface.

The most famous optical illusion, known as Müller-Lyer's, consists of two lines of equal length with extensions pointing in opposite directions, as shown in Fig. 179. When cardboard cutouts of these forms were pressed against the side of the thigh or the abdomen, most observers reported that the length of the line with extensions pointing away from the center seemed longer than the line with extensions pointing inward. Several tests were necessary to find the best length of the different sections, which were 1 mm thick. It was found most convenient to give the center line a length of about 14 cm and to make the extensions about 7 cm long. Care had to be taken not to press too hard on the skin surface, for it is mainly the superficial part of the skin and not the deep underlying muscle layer that is affected.

I consider this apparent change in the length of the center line to be produced in some degree by lateral inhibition. If an angular pattern, such as the solid line in *a* of Fig. 180 is pressed

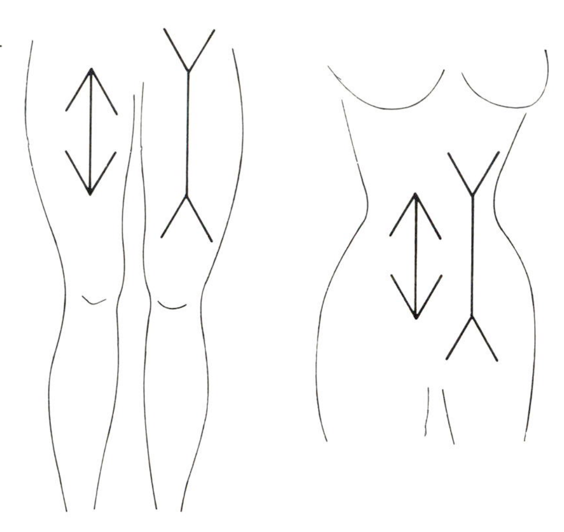

Fig. 179. Well-known optical ilusions represented on the skin.

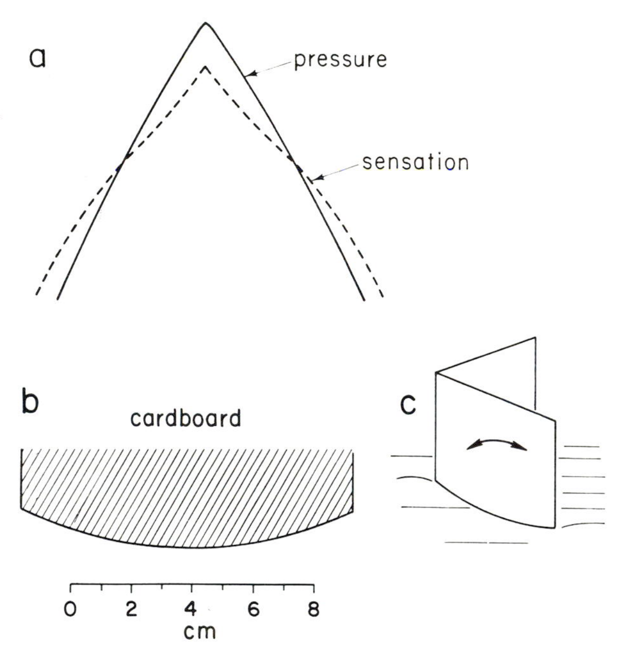

Fig. 180. A cardboard stimulus with a rounded edge as shown at *b* was applied to the skin with a rocking motion as at *c*. The observer perceived the angle of the form as more obtuse than the actual pressure, as shown at *a*.

against the surface of the skin, the pattern of sensation is that of the broken line. And it is this apparent change in the angle that, in the Müller-Lyer pattern, produces the shortening or lengthening of the center line. At the apex of the wedge the distance between the two pieces of cardboard is small and (as earlier observations show) we are unable to discriminate the difference between right and left edges. They produce a common pressure sensation in the middle. As the distance between corresponding points of the wedge increases, a distance is reached where the corresponding points inhibit one another. This point of inhibition is not recognized as such but only as a decrease in the magnitude of sensation along the sides of the wedge. For even larger distances, however, the pressure sensation in the middle of the wedge is replaced by two separate sensations of the two edges. The sudden change from a summation to two separate sensations produces the illusion that the angle between the two pieces of cardboard is larger than it really is.

Pressure on the surface of the skin produces rapid adaptation, and the cardboard experiment is carried out in a more satisfactory manner if the whole cardboard is not pressed on the skin all at once. By the use of a curved edge as seen in *b* of Fig. 180, the card may be rolled slowly along the skin surface. For this purpose the curvature needs to be only about 10 to 20°.

This is only a first attempt to show the possibility of inhibitory phenomena of the skin sense organs. The displacement of sensory locations in the manner described suggests effects similar to those of optical illusions.

From Fig. 181 we may conclude that the lengthening of a line will be large when the end of the line is attached to the tip of the wedge, as in *c* of Fig. 182. The shortening will be smaller when the tip of the wedge forms the end of the line, as in *a* of Fig. 182. From this relation it appears that a comparison of part *b* of Fig. 182 with *a* and *c* should show that the difference between the vertical lines of *c* and *b* should exceed that between *b* and *a*.

Several optical illusions besides that of Müller-Lyer can easily be duplicated on the skin. An example is shown in Fig. 183 (*a* and *b*), where the distance between two lines is compared when the space between them is empty and when it contains other lines. This is the Helmholtz illusion (1856), in which *a* seems greater than *b*. A second effect appears when between the outer lines two others are placed only a short distance away, as in *c* of this figure. This is the Delboeuf illusion (1865), and now the apparent distance between the outer lines seems greater for *b* than for *c*.

These illusions are readily produced on the skin, as shown in Fig. 184, where a device is used to reproduce the Delboeuf illusion on the inner surface of the palm. Four pieces of cardboard with rounded edges were attached to wooden blocks as shown. Rotating the block 180° gave a comparison of the two forms. The perceived distance was clearly greater when only two edges were pressed against the palm. Care was taken to avoid adaptation by applying the stimuli only briefly.

The Delboeuf illusion may be explained as a flowing together of the effect of two adjacent stimuli to a central position.

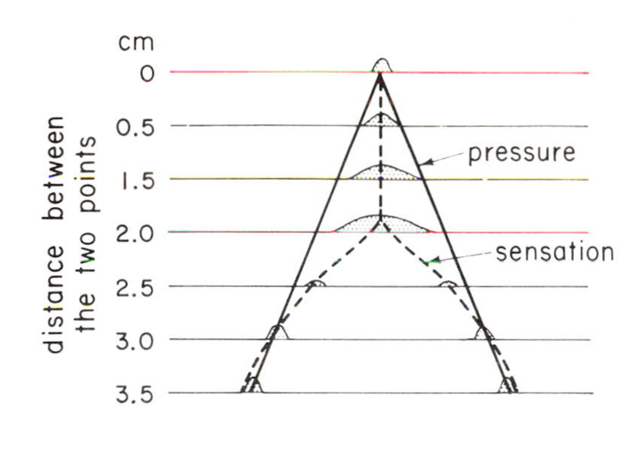

Fig. 181. A theory of the Müller-Lyer illusion on the skin.

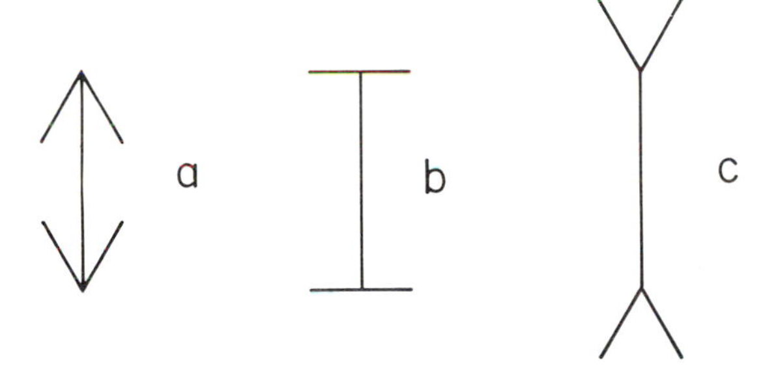

Fig. 182. Parts of the Müller-Lyer illusion with an intermediate form, *b*.

Fig. 183. The Helmholtz and Delboeuf illusions.

Fig. 184. Method of studying the Delboeuf illusion on the palm.

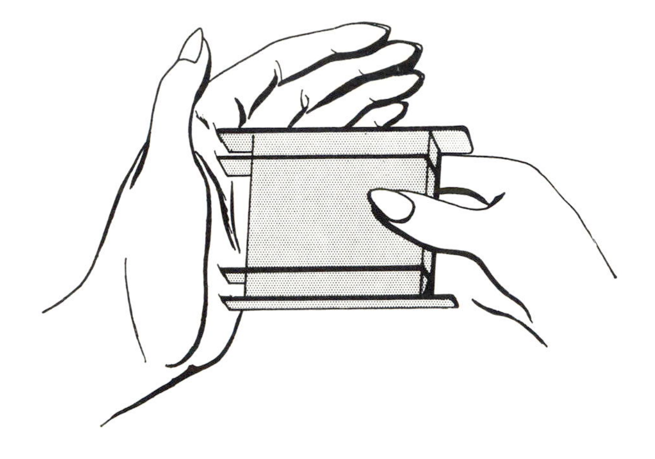

The use of inhibitory localization to indicate the similarity of two stimuli

If we present simultaneously a click stimulus to the right ear, and another that is slightly different to the left ear, they will usually add their effects to produce a sharply localized image. If the two clicks are carefully matched in loudness, this image changes its position when a time delay is introduced for one of them. As the qualitative difference between the two clicks increases, then in general the lateral spread of the sound image increases also. However, there is a limit to the qualitative difference that may be introduced, beyond which the clicks no longer add together to one image but fall apart so that each click is heard separately in its proper ear. The two situations are pictured in Fig. 185, where *A* represents a small qualitative difference and *B* a larger one.

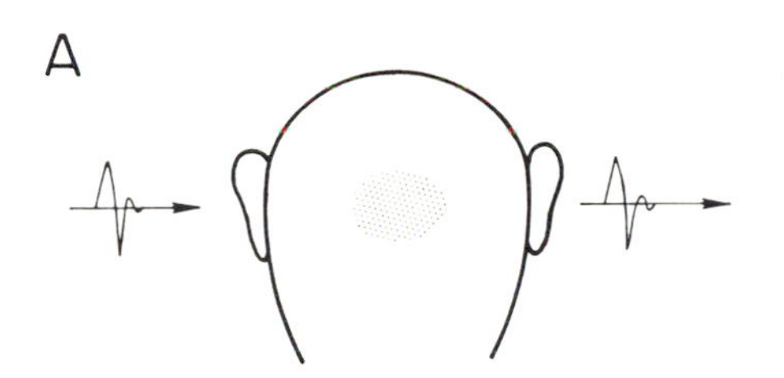

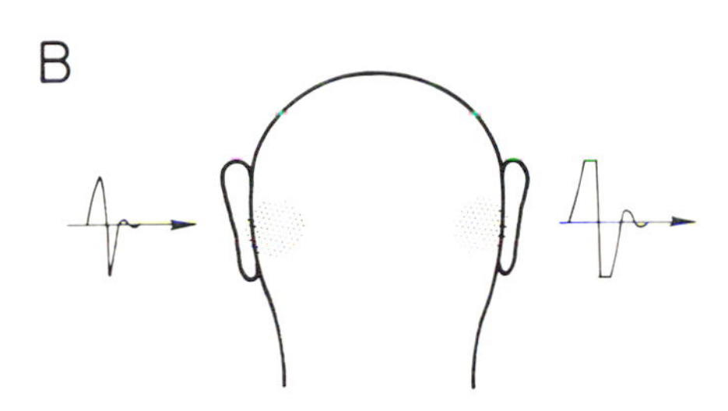

Fig. 185. Localization as a method of evaluating a difference between two stimuli. From *Science* [115].

This phenomenon can be used to discover the variables that cause the two clicks to separate. It is an interesting field of investigation to ascertain the difference in the frequency band of the click that is required to produce a separation. This type of observation shows that for localization it is the onset of the click that is important, and a large difference in the following decay time does not have a tendency to separate the clicks.

The same sort of investigation can be made with other forms of distortion, such as phase distortion, and it may seem surprising how small the phase differences are that can lead to a separation of the two clicks. On the other hand, nonlinear distortion mainly affects the loudness and does not contribute to the separation. By the use of this method it is possible to a certain extent to obtain a rank order of types of distortion according to their importance. This may be done by increasing the distortion (for equally loud stimuli) to such an extent that the sound image begins to break into two. There is probably a correspondence between this method of estimating the importance of various forms of distortion and some of the speech recognition tests.

This same method can be used to investigate the importance of onset envelopes of tones that are nearly pure. For this purpose use could be made of tone pulses that have the same frequency but different onset and decay times, with observation of the magnitude of the difference that causes the sound image to begin to separate into two. The experiment could be carried out by changing the onset and decay envelopes gradually and by changing the frequency of one tone pulse relative to the other. There is a certain frequency difference at which the sound image falls apart. This frequency difference will be altered with variations in loudness and by the use of different kinds of onset and decay.

The method can be used also to determine the relative magnitudes of common features of two different sensations. It can even be employed to detect common properties between hearing and vibratory sensations by trying to produce changes of localization by means of a time delay.

Perhaps the most successful attempt with this method was an investigation of the four basic taste qualities, so as to discover a closer relation between certain pairs of them. Thus there is evidence for a closer relation between bitter and sweet and between salty and sour than between other combinations like bitter and salty or sweet and sour. To test this assumption, the taste stimulating equipment described earlier (Fig. 83) was used and an investigation made of the changes in localization of a taste sensation caused by time delays. In the present instance a bitter stimulus was applied to one side of the tongue and a sweet stimulus to the other side. Both stimuli were introduced at the same moment and they were adjusted to give equal magnitudes of sensation by the use of the mixing device shown in Fig. 5. Each lasted 1 second. The experiments showed that bitter and sweet can produce a unitary sensation in the middle of the tongue and that this sensation can be moved from one side to the other by the use of time delays. There is a certain relationship between bitter and sweet. The same relationship seems to hold for salty and sour. However, other combinations of the four tastes, such as bitter and salty or sweet and sour did not give a localization in the middle. These last were located separately on the two sides of the tongue, at the openings of the stimulating block. It appears that there is only a limited relation between sweet and sour and between bitter and salty.

The same experiment was repeated by using warm or cold water on one side of the tongue and one of the basic taste qualities on the other side. After careful adjustment of the magnitudes of the sensations for equality when the stimuli were presented separately, it was found that warm seems to interact somewhat with bitter or with sweet, whereas cold seems to interact with salty and with sour. These interactions are represented in Fig. 186.

The summation effect between certain taste

Fig. 186. Degrees of resemblance among the four taste qualities and warm and cold as determined by the localization method. From *Science* [115].

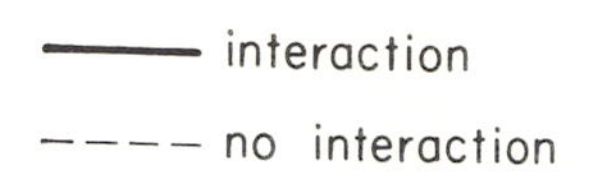

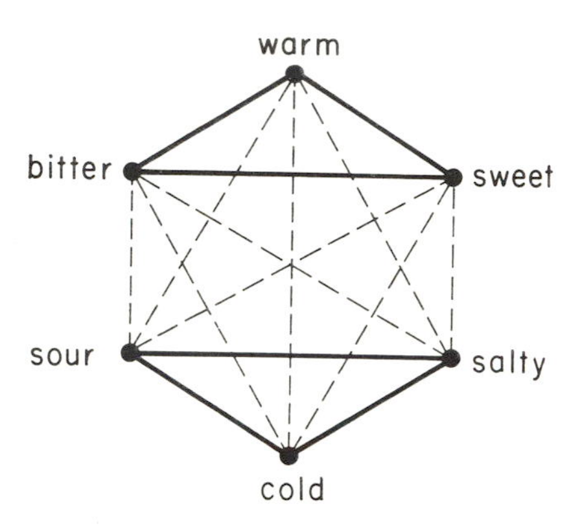

qualities should be found for the electrical stimulation of single papillae. For this purpose an elaborate setup was designed to stimulate a papilla with brief electric pulses. Then two papillae were selected that plainly produced a salty taste. The electrical arrangement is shown in Fig. 187. The two potentiometers were adjusted to give the same magnitudes of sensation when the two papillae were stimulated through the point electrodes. This adjustment of magnitude is difficult because the sensation varies rapidly as a function of stimulating voltage. If the adjustment is properly carried out it is possible to switch on the current to both electrodes simultaneously and have the sensations appear in the middle between the two electrodes as shown in *a* of Fig. 188. On the introduction of a time delay at one electrode the locus of an electrically produced taste will change in a manner indicated by the figure. In these observations a large

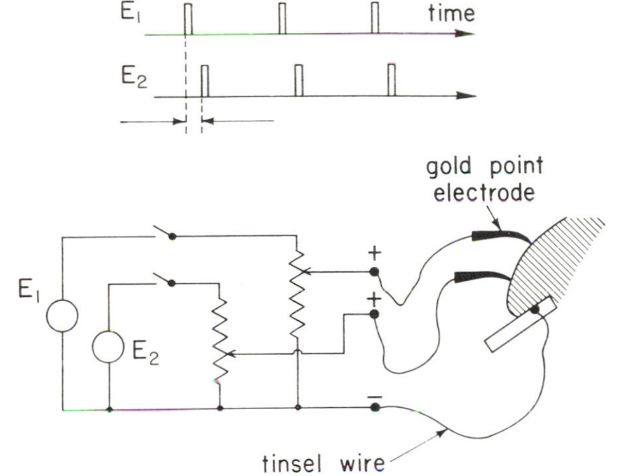

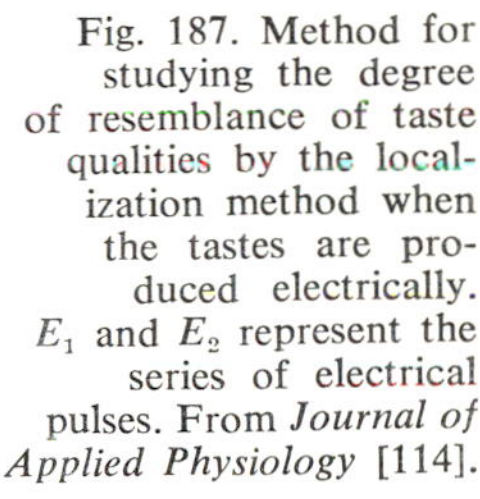
Fig. 187. Method for studying the degree of resemblance of taste qualities by the localization method when the tastes are produced electrically. E_1 and E_2 represent the series of electrical pulses. From *Journal of Applied Physiology* [114].

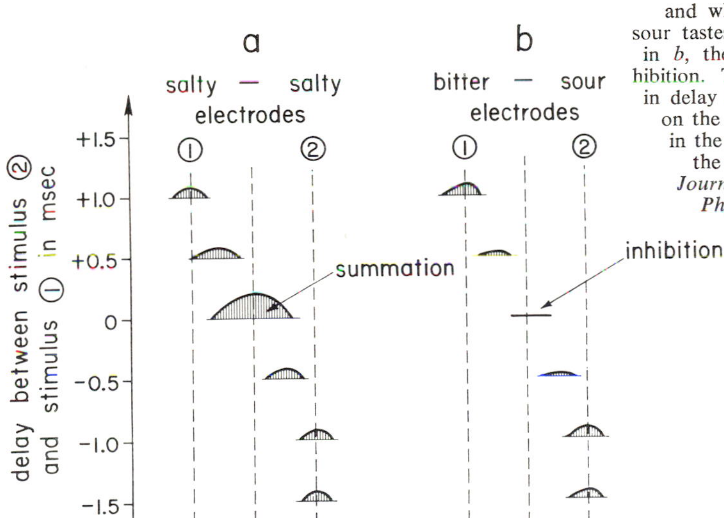

Fig. 188. Results of comparing two pairs of tastes by the foregoing method. When two salty tastes are used, as in *a*, the result of simultaneous presentation is summation, and when bitter and sour tastes are used, as in *b*, the result is inhibition. The variations in delay as represented on the ordinate assist in the recognition of the effects. From *Journal of Applied Physiology* [114].

indifferent electrode was placed on the side of the tongue and the point electrodes were always made positive, for only positive voltage produces a taste sensation for energies near the threshold. On the tip of the tongue it is possible in most observers to find some taste papillae that on stimulation give a bitter taste similar to that produced by a quinine solution. If one electrode is placed on this papilla and another on a papilla giving a sour taste, there is no combination of bitter and sour for zero time difference, but the bitter and sour tastes remain separate, each beneath its own stimulating electrode. This observation indicates that the form of interaction between bitter and sour is different from that between two papillae giving the same taste (e.g., salty).

There are indications that this method of comparing two stimuli provides a hint also about the strength of their interaction.

In a final word on the subject of inhibition I wish to point out once more that inhibition plays its important role as developed here mainly within the range of medium and large magnitudes of sensation. Near their thresholds the sensations more often display summational effects, and these seem to dominate the picture. However, a role for inhibition even for threshold sensations ought not to be dismissed entirely.

AUTHOR'S BIBLIOGRAPHY

1. Über den Einfluss der nichtlinearen Eisenverzerrungen auf die Güte und Verständlichkeit eines Telephonie-Übertragungssystemes, *Elektrische Nachrichten-technik*, 1928, 5, 231-246.
2. Zur Theorie des Hörens; Die Schwingungsform der Basilarmembran, *Physikalische Zeitschrift*, 1928, 29, 793-810.
3. Zur Theorie des Hörens; Über die Bestimmung des einem reinen Tonempfinden entsprechenden Erregungsgebietes der Basilarmembran vermittelst Ermüdungserscheinungen, *Physikalische Zeitschrift*, 1929, 30, 115-125.
4. Zur Theorie des Hörens; Über die eben merkbare Amplituden- und Frequenzänderung eines Tones; Die Theorie der Schwebungen, *Physikalische Zeitschrift*, 1929, 30, 721-745.
5. Zur Theorie des Hörens; Über das Richtungshören bei einer Zeitdifferenz oder Lautstärkenungleichheit der beiderseitigen Schalleinwirkungen, *Physikalische Zeitschrift*, 1930, 31, 824-835, 857-868.
6. Über das Fechnersche Gesetz und seine Bedeutung für die Theorie der akustischen Beobachtungsfehler und die Theorie des Hörens, *Annalen der Physik*, series 5, 1930, 7, 329-359.
7. Sur la théorie de l'audition, *L'Année Psychologique*, 1930, 31, 63-96.
8. Bemerkungen zur Theorie der günstigsten Nachhalldauer von Räumen, *Annalen der Physik*, series 5, 1931, 8, 851-873.
9. Über die Messung der Schwingungsamplitude fester Körper, *Annalen der Physik*, series 5, 1931, 11, 227-232.

10. Über die Ausbreitung der Schallwellen in anisotropen dünnen Platten, *Zeitschrift für Physik*, 1932, 79, 668-671.
11. Zur Theorie des Hörens bei der Schallaufnahme durch Knochenleitung, *Annalen der Physik*, series 5, 1932, 13, 111-136.
12. Über den Einfluss der durch den Kopf und den Gehörgang bewirkten Schallfeldverzerrungen auf die Hörschwelle, *Annalen der Physik*, series 5, 1932, 14, 51-56.
13. Über die Schallfeldverzerrungen in der Nähe von absorbierenden Flächen und ihre Bedeutung für die Raumakustik, *Zeitschrift für technische Physik*, 1933, 14, 6-10.
14. Über die Hörsamkeit der Ein- und Ausschwingvorgänge mit Berücksichtigung der Raumakustik, *Annalen der Physik*, series 5, 1933, 16, 844-860.
15. Über den Knall und die Theorie des Hörens, *Physikalische Zeitschrift*, 1933, 34, 577-582.
16. Über die Hörsamkeit kleiner Musikräume, *Annalen der Physik*, series 5, 1934, 19, 665-679.
17. Über die nichtlinearen Verzerrungen des Ohres, *Annalen der Physik*, series 5, 1934, 20, 809-827.
18. Über die Hörsamkeit von Konzert- und Rundfunksälen, *Elektrische Nachrichten-technik*, 1934, 11, 369-375.
19. Physikalische Problem der Hörphysiologie, *Elektrische Nachrichten-technik*, 1935, 12, 71-83.
20. Über akustische Reizung des Vestibularapparates, *Pflügers Archiv für die gesamte Physiologie*, 1935, 236, 59-76.
21. Über akustische Rauhigkeit, *Zeitschrift für technische Physik*, 1935, 16, 276-282.
22. Über die Herstellung und Messung langsamer sinusförmiger Luftdruckschwankungen, *Annalen der Physik*, series 5, 1936, 25, 413-432.

23. Über die Hörschwelle und Fühlgrenze langsamer sinusförmiger Luftdruckschwankungen, *Annalen der Physik*, series 5, 1936, 26, 554-566.
24. Fortschritte der Hörphysiologie, *Zeitschrift für technische Physik*, 1936, 17, 522-528.
25. Zur Physik des Mittelohres und über das Hören bei fehlerhaftem Trommelfell, *Akustische Zeitschrift*, 1936, 1, 13-23.
26. Über die photoelektrische Fourier-Analyse eines gegebenen Kurvenzuges, *Elektrische Nachrichten-technik*, 1937, 14, 157-161.
27. Über subjektive harmonische Teiltöne (Antwort zu Bemerkungen des Herrn Lottermoser), *Akustische Zeitschrift*, 1937, 2, 149.
28. Über die mechanische Frequenzanalyse einmaliger Schwingungsvorgänge und die Bestimmung der Frequenzabhängigkeit von Übertragungssystemen und Impedanzen mittels Ausgleichsvorgängen, *Akustische Zeitschrift*, 1937, 2, 217-224.
29. Über die Entstehung der Entfernungsempfindung beim Hören, *Akustische Zeitschrift*, 1938, 3, 21-31.
30. Psychologie und Fernsprechtechnik, *Forschungen und Fortschritte*, 1938, 14, 342-344.
31. Über die piezoelektrische Messung der absoluten Hörschwelle bei Knochenleitung, *Akustische Zeitschrift*, 1939, 4, 113-125.
32. Über die mechanische-akustischen Vorgänge beim Hören, *Acta Oto-laryngologica*, 1939, 27, 281-296, 388-396.
33. Über die Vibrationsempfindung, *Akustische Zeitschrift*, 1939, 4, 316-334.
34. Über die Empfindlichkeit des stehenden und sitzenden Menschen gegen sinusförmige Erschütterungen, *Akustische Zeitschrift*, 1939, 4, 360-369.
35. Über die Stärke der Vibrationsempfindung und ihre objektive Messung, *Akustische Zeitschrift*, 1940, 5, 113-124.

36. Über die Sicherheit und Reproduzierbarkeit künstlerischer Urteile beim Rundfunk, *Arbeitstagung der deutschen Fachbeiräte für Sing- und Sprechkultur*, Wien, April 1940.
37. The neural terminations responding to stimulation of pressure and vibration, *Journal of Experimental Psychology*, 1940, 26, 514-519.
38. Über die Messung der Schwingungsamplitude der Gehörknöchelchen mittels einer kapazitiven Sonde, *Akustische Zeitschrift*, 1941, 6, 1-16.
39. Über die Elastizität der Schneckentrennwand des Ohres, *Akustische Zeitschrift*, 1941, 6, 265-278; On the elasticity of the cochlear partition, *Journal of the Acoustical Society of America*, 1948, 20, 227-241.
40. Über die Schallausbreitung bei Knochenleitung, *Zeitschrift für Hals- Nasen- und Ohrenheilkunde*, 1941, 47, 430-442.
41. Über das Hören der eigenen Stimme, *Anzeiger der Akademie der Wissenschaften in Wien, math-nat. Cl.*, 1941, 78, 61-70.
42. Über die Schwingungen der Schneckentrennwand beim Präparat und Ohrenmodell, *Akustische Zeitschrift*, 1942, 7, 173-186; The vibration of the cochlear partition in anatomical preparations and in models of the inner ear, *Journal of the Acoustical Society of America*, 1949, 21, 233-245.
43. Über die Resonanzkurve und die Abklingzeit der verschiedenen Stellen der Schneckentrennwand, *Akustische Zeitschrift*, 1943, 8, 66-76; On the resonance curve and the decay period at various points on the cochlear partition, *Journal of the Acoustical Society of America*, 1949, 21, 245-254.
44. Über die direkte mikroskopische Ausmessung der Resonanzschärfe und Dämpfung der sogenannten Ohrenresonatoren, *Forschungen und Fortschritte*, 1943, 19, 364-365.

45. Über die mechanische Frequenzanalyse in der Schnecke verschiedener Tiere, *Akustische Zeitschrift*, 1944, 8, 3-11.
46. Über die Frequenzauflösung in der menschlichen Schnecke, *Acta Oto-laryngologica*, 1944, 32, 60-84.
47. The variation of phase along the basilar membrane with sinusoidal vibrations, *Journal of the Acoustical Society of America*, 1947, 19, 452-460.
48. The sound pressure difference between the round and the oval windows and the artificial window of labyrinthine fenestration, *Acta Oto-laryngologica*, 1947, 35, 301-315.
49. A new audiometer, *Acta Oto-laryngologica*, 1947, 35, 411-422; and *Archiv für elektrischer Übertragung*, 1947, 1, 13-16.
50. The recruitment phenomenon and difference limen in hearing and vibration sense, *Laryngoscope*, 1947, 57, 765.
51. (With W. A. Rosenblith) The early history of hearing–observations and theories, *Journal of the Acoustical Society of America*, 1948, 20, 727-748.
52. Vibration of the head in a sound field, and its role in hearing by bone conduction, *Journal of the Acoustical Society of America*, 1948, 20, 749-760.
53. The structure of the middle ear and the hearing of one's own voice by bone conduction, *Journal of the Acoustical Society of America*, 1949, 21, 217-232.
54. Über die Mondillusion, *Experientia*, 1949, 5, 326.
55. The moon illusion and similar auditory phenomena, *American Journal of Psychology*, 1949, 62, 540-552.
56. Interchangeable pencil-type micromanipulator, *Science*, 1950, 111, 667-669.
57. Suggestions for determining the mobility of the stapes by means of an endotoscope for the middle ear, *Laryngoscope*, 1950, 60, 97-110.
58. D-C potentials and energy balance of the cochlear

partition, *Journal of the Acoustical Society of America*, 1951, 23, 576-582.

59. (With W. A. Rosenblith) The mechanical properties of the ear, Chapter 27 in *Handbook of Experimental Psychology* (S. S. Stevens, ed.), New York, John Wiley and Sons, 1951, pp. 1075-1115.
60. Microphonics produced by touching the cochlear partition with a vibrating electrode, *Journal of the Acoustical Society of America*, 1951, 23, 29-35.
61. The coarse pattern of the electrical resistance in the cochlea of the guinea pig (electro-anatomy of the cochlea), *Journal of the Acoustical Society of America*, 1951, 23, 18-28.
62. Resting potentials inside the cochlear partition of the guinea pig, *Nature*, 1952, 169, 241-242.
63. Micromanipulator with four degrees of freedom, *Transactions of the American Microscopical Society*, 1952, 71, 306-310.
64. D-C resting potentials inside the cochlear partition, *Journal of the Acoustical Society of America*, 1952, 24, 72-76.
65. Gross localization of the place of origin of the cochlear microphonics, *Journal of the Acoustical Society of America*, 1952, 24, 399-409.
66. Direct observation of the vibrations of the cochlear partition under a microscope, *Acta Oto-laryngologica*, 1952, 42, 197-201.
67. Description of some mechanical properties of the organ of Corti, *Journal of the Acoustical Society of America*, 1953, 25, 770-785.
68. Shearing microphonics produced by vibrations near the inner and outer hair cells, *Journal of the Acoustical Society of America*, 1953, 25, 786-790.
69. Note on the definition of the term: hearing by bone conduction, *Journal of the Acoustical Society of America*, 1954, 26, 106-107.

70. (With E. G. Wever and M. Lawrence) A note on recent developments in auditory theory, *Proceedings of the National Academy of Sciences*, Washington, 1954, 40, 508-512.
71. Some electro-mechanical properties of the organ of Corti, *Annals of Otology, Rhinology, and Laryngology*, 1954, 63, 448-468.
72. Subjective cupulometry, *Archives of Otolaryngology*, 1955, 61, 16-28.
73. Paradoxical direction of wave travel along the cochlear partition, *Journal of the Acoustical Society of America*, 1955, 27, 137-145.
74. Human skin perception of traveling waves similar to those on the cochlea, *Journal of the Acoustical Society of America*, 1955, 27, 830-841.
75. Beitrag zur Frage der Frequenzanalyse in der Schnecke, *Archiv für Ohren- Nasen- Kehlkopfheilkunde und Zeitschrift für Hals- Nasen- Ohrenheilkunde*, 1955, 167, 238-255.
76. Current status of theories of hearing, *Science*, 1956, 123, 779-783.
77. Preparatory and air-driven micromanipulators for electrophysiology, *Review of Scientific Instruments*, 1956, 27, 690-692.
78. Simplified model to demonstrate the energy flow and formation of traveling waves similar to those found in the cochlea, *Proceedings of the National Academy of Sciences*, Washington, 1956, 42, 930-944.
79. Sensations on the skin similar to directional hearing, beats, and harmonics of the ear, *Journal of the Acoustical Society of America*, 1957, 29, 489-501.
80. The ear, *Scientific American*, 1957, 197, 66-78.
81. Neural volleys and the similarity between some sensations produced by tones and by skin vibrations, *Journal of the Acoustical Society of America*, 1957, 29, 1059-1069.

82. Pendulums, traveling waves, and the cochlea: introduction and script for a motion picture. *Laryngoscope*, 1958, 68, 317-327.
83. Funneling in the nervous system and its role in loudness and sensation intensity on the skin, *Journal of the Acoustical Society of America*, 1958, 30, 399-412.
84. Similarities between hearing and skin sensations, *Psychological Review*, 1959, 66, 1-22.
85. Synchronism of neural discharges and their demultiplication in pitch perception on the skin and in hearing, *Journal of the Acoustical Society of America*, 1959, 31, 338-349.
86. (With J. Lempert) Improvement of sound transmission in the fenestrated ear by the use of shearing forces, *Laryngoscope*, 1959, 69, 876-883.
87. Neural funneling along the skin and between the inner and outer hair cells of the cochlea, *Journal of the Acoustical Society of America*, 1959, 31, 1236-1249.
88. *Experiments in Hearing*, New York, McGraw-Hill, 1960, 745 pp.
89. Über die Gleichartigkeit einiger nervöser Prozesse beim Hören und Vibrationssinn, in *Proceedings of the Third International Congress on Acoustics*, Amsterdam, Elsevier, 1961, pp. 13-20.
90. Neural inhibitory units of the eye and skin: Quantitative description of contrast phenomena, *Journal of the Optical Society of America*, 1960, 50, 1060-1070.
91. Pitch sensation and its relation to the periodicity of the stimulus; hearing and skin vibrations, *Journal of the Acoustical Society of America*, 1961, 33, 341-348.
92. Experimental models of the cochlea with and without nerve supply, in *Neural Mechanisms of the Auditory and Vestibular Systems* (G. L. Rasmussen and W. F. Windle, eds.), Springfield, Ill., Charles C Thomas, 1960, pp. 3-20.

93. The influence of inhibition on the sensation pattern of the skin and the eye, in *Symposium on Cutaneous Sensitivity*, Fort Knox, Ky., U.S. Army Medical Research Laboratory, Report No. 424, 1960, pp. 50-62.
94. Are surgical experiments on human subjects necessary? *Laryngoscope*, 1961, 71, 367-376.
95. Concerning the fundamental component of periodic pulse patterns and modulated vibrations observed on the cochlear model with nerve supply, *Journal of the Acoustical Society of America*, 1961, 33, 888-896.
96. (With E. G. Wever, W. E. Rahm, Jr., and J.H.T. Rambo) A new method of perfusion for the fixation of tissues, *Laryngoscope*, 1961, 71, 1534-1547.
97. Abweichungen vom Ohmschen Gesetz der Frequenzauflösung beim Hören, *Akustische Beihefte, Acustica,* 1961, 11, 241-244.
98. The gap between the hearing of external and internal sounds, in *Biological Receptor Mechanisms*, Cambridge, England, University Press, 1962, pp. 267-288 (Symposium No. 16, Society for Experimental Biology).
99. Comments on the measurement of the relative size of dc potentials and microphonics in the cochlea, *Journal of the Acoustical Society of America*, 1962, 34, 124.
100. Can we feel the nervous discharges of the end organs during vibratory stimulation of the skin? *Journal of the Acoustical Society of America*, 1962, 34, 850-856.
101. Concerning the pleasures of observing, and the mechanics of the inner ear, in *Les Prix Nobel en 1961*, Stockholm, Imprimerie Royale P. A. Norstedt & Söner, 1962, pp. 184-208.
102. Synchrony between nervous discharges and periodic stimuli in hearing and on the skin, *Annals of Otology, Rhinology, and Laryngology*, 1962, 71, 678-692.
103. Wave motion in an inhomogeneous system: The

movements of the basilar membrane in the cochlea, Script of a motion picture film, May 1962.

104. Lateral inhibition of heat sensations on the skin, *Journal of Applied Physiology*, 1962, 17, 1003-1008.

105. Hearing theories and complex sounds, *Journal of the Acoustical Society of America*, 1963, 35, 588-601.

106. Three experiments concerned with pitch perception, *Journal of the Acoustical Society of America*, 1963, 35, 602-606.

107. Letter to editor and contributors of the Békésy commemorative issue of the Journal of the Acoustical Society of America, September 1962, *Journal of the Acoustical Society of America*, 1963, 35, 120.

108. Concluding remarks of the round table discussion on the frequency analysis of the normal and pathological ear (Sixth International Congress of Audiology, Copenhagen, September 1962), *International Audiology*, 1963, 2, 26-29.

109. Interaction of paired sensory stimuli and conduction in peripheral nerves, *Journal of Applied Physiology*, 1963, 18, 1276-1284.

110. Modification of sensory localization as a consequence of oxygen intake and reduced blood flow, *Journal of the Acoustical Society of America*, 1963, 35, 1183-1187.

111. Rhythmical variations accompanying gustatory stimulation observed by means of localization phenomena, *Journal of General Physiology*, 1964, 47, 809-825.

112. Olfactory analogue to directional hearing, *Journal of Applied Physiology*, 1964, 19, 369-373.

113. Die gegenseitige Hemmung von Sinnesreizen bei kleinen Zeitdifferenzen, *Naturwissenschaftliche Rundschau*, 1964, 17, 209-216.

114. Sweetness produced electrically on the tongue and its relation to taste theories, *Journal of Applied Physiology*, 1964, 19, 1105-1113.

115. Duplexity theory of taste, *Science*, 1964, 145, 834-835.
116. The effect of adaptation on the taste threshold observed with a semiautomatic gustometer, *Journal of General Physiology*, 1965, 48, 481-488.
117. Cochlear mechanics, in *Theoretical and Mathematical Biology* (T. H. Waterman and H. J. Morowitz, eds.), New York, Blaisdell, 1965, pp. 172-197.
118. Inhibition and the time and spatial patterns of neural activity in sensory perception, *Annals of Otology, Rhinology, and Laryngology*, 1965, 74, 445-462.
119. Temperature coefficient of electrical thresholds of taste sensations, *Journal of General Physiology*, 1965, 49, 27-35.
120. Taste theories and the chemical stimulation of single papillae, *Journal of Applied Physiology*, 1966, 21, 1-9.
121. A large mechanical model of the cochlea with nerve supply (in press).
122. Mach band type lateral inhibition in different sense organs, to appear in *Journal of General Physiology*.

REFERENCES

Adrian, E. D., *The Basis of Sensation: The Action of the Sense Organs*, New York, W. W. Norton and Co., 1928.

Arvanitaki, A., Recherches sur la réponse oscillatoire locale de l'axone géant isolé de "Sepia," *Archives Internationales de Physiologie*, 1939, 49, 209.

Baumgartner, G., and Hakas, P., Reaktionen einzelner Opticusneuronen und corticaler Nervenzellen der Katze im Hell-Dunkel-Grenzfeld. Simultankontrast, *Pflügers Archiv für die gesamte Physiologie*, 1959, 270, 29.

Beranek, L. L., *Music, Acoustics and Architecture*, New York, London, John Wiley & Sons, 1962.

Bujas, Z. and Ostojcit, A., L'évolution de la sensation gustative en fonction du temps d'excitation, *Acta Instituti Psychologici Universitatis Zagrebensis*, 1939, 3, 24 pp.

Burkhardt, D., Allgemeine Sinnesphysiologie und Elektrophysiologie der Receptoren, *Fortschritte der Zoologie*, 1961, 13, 146-189.

Charpentier, A., Réaction oscillatoire de la rétine sous l'influence des excitations lumineuses, *Archives de Physiologie Normale et Pathologique*, series 5, 1892, 4, 541-553.

Delboeuf, J., (1) Note sur certaines illusions d'optique; essai d'une théorie psychophysique de la manière dont l'oeil apprécie les distances et les angles, *Académie Royale des Sciences, des Lettres et des Beaux-Arts de Belgique, Brussels, Bulletins*, series 2, 1865, 19, 195-216. (2) Seconde note sur de nouvelles illusions d'optique; Essai d'une théorie psychophysique de la manière dont l'oeil apprécie les grandeurs, *ibid.*, 20, 70-97.

Descartes, René, *L'Homme*, 2nd edn., Paris, Charles Angot, 1677, esp. pp. 15-18.

Diestel, H. G., Akustische Messungen an einem mechanischen Modell des Innenohres, *Acustica*, 1954, 4, 489-499.

Dijkgraaf, S., Functioning and significance of the lateral-line organs, *Biological Review*, 1963, 38, 51-105.

Eccles, J. C., *The Physiology of Nerve Cells*, Baltimore, Johns Hopkins Press, 1957.

Erlanger, J., and Gasser, H. S., *Electrical Signs of Nervous Activity*, Philadelphia, University of Pennsylvania Press, 1937.

Field, John, ed., *Handbook of Physiology*, Section 1, *Neurophysiology*, 1, Washington, D.C., American Physiological Society, 1959.

Florey, Ernst, ed., Nervous Inhibition: *Proceedings of the Second International Symposium on Nervous Inhibition, Friday Harbor*, New York, Pergamon Press, 1961.

Frey, M. von, (1) Die Tangoreceptoren des Menschen, *in* A. Bethe, ed., *Handbuch der normalen und pathologischen Physiologie*, 11, *Receptionsorgane*, 1, Berlin, Springer-Verlag, 1926, pp. 94-130, esp. p. 97. (2) Die Haut als Sinnesfläche, *in* J. Jadassohn, ed., *Handbuch der Haut- und Geschlechtskrankheiten*, Berlin, Springer-Verlag, 1929, vol. 1, part 2, pp. 91-160, esp. pp. 111-138.

Frey, M. von, and Goldman, A., Der zeitliche Verlauf der Einstellung bei den Druckempfindungen, *Zeitschrift für Biologie*, 1914, 65, 183.

Fröhlich, F. W., (1) Beiträge zur allgemeinen Physiologie der Sinnesorgane, *Zeitschrift für Sinnesphysiologie*, 1913, 48, 28. (2) Über oszillierende Erregungsvorgänge im Sehfeld, *Zeitschrift für Sinnesphysiologie*, 1921, 52, 52-59.

Gray, Henry, *Anatomy of the Human Body*, 27th edn., edited by C. M. Goss, Philadelphia, Lea & Febiger, 1959, p. 1112 and Fig. 956.

Haas, Helmut, Über den Einfluss eines Einfachechos auf die Hörsamkeit von Sprache, *Acustica*, 1951, 1, 49-58.

Hardy, Mary, Observations on the innervation of the

macula sacculi in man, *Anatomical Record*, 1934, 59, 403-418.

Hartline, H. K., Inhibition of activity of visual receptors by illuminating nearby retinal areas in the Limulus eye, *Federation Proceedings*, 1949, 8, 69.

Hartline, H. K., Ratliff, F., and Miller, W. H., (1) Neural interaction in the eye and the integration of receptor activity, *Annals of the New York Academy of Sciences*, 1958, 74, 210-222. (2) Inhibitory interaction in the retina and its significance in vision, *in* E. Florey, ed., *Nervous Inhibition*, London, Pergamon, 1961, pp. 241-284.

Hartline, H. K., Wagner, H. G., and Ratliff, F., Inhibition in the eye of *limulus*, *Journal of General Physiology*, 1956, 39, 651-673.

Held, Hans, Die Cochlea der Säuger und der Vögel, ihre Entwicklung und ihr Bau, *in* A. Bethe, ed., *Handbuch der normalen und pathologischen Physiologie*, 11, *Receptionsorgane* 1, Berlin, Springer-Verlag, 1926, pp. 467-543, esp. p. 503, Fig. 105.

Helmholtz, H. von, (1) Messungen über den zeitlichen Verlauf der Zuckung animalischer Muskeln und die Fortpflanzungsgeschwindigkeit der Reizung in den Nerven, *Archiv für Anatomie und Physiologie, Physiologische Abtheilung*, 1850, 276-364. (2) Messungen über Fortpflanzungsgeschwindigkeit der Reizung in den Nerven, *Archiv für Anatomie und Physiologie, Physiologische Abtheilung*, series 2, 1852, 199-216. (3) *Handbuch der physiologischen Optik*, Leipzig, 3 vols., 1856-1866.

Hering, E., *Grundzüge der Lehre vom Lichtsinn*, Berlin, Springer-Verlag, 1920.

Hoagland, H., Impulses from sensory nerves of catfish, *Proceedings of the National Academy of Sciences*, Washington, 1932, 18, 701-705.

Hoffmann, P., Über die doppelte Innervation der Krebsmuskeln; Zugleich ein Beitrag sur Kenntnis nervöser

Hemmungen, *Zeitschrift für Biologie*, 1914, 63, 411-442.

Holst, Erich von, and Mittelstaedt, Horst, Das Reafferenzprinzip (Wechselwirkungen zwischen Zentralnervensystem und Peripherie), *Naturwissenschaften*, 1950, 37, 464-476.

Hornbostel, E. M. von, (1) Beobachtungen über ein- und zweiohriges Hören, *Psychologische Forschung*, 1923, 4, 64-114. (2) Das raümliche Hören, *in* A. Bethe, ed., *Handbuch der normalen und pathologischen Physiologie*, 11, *Receptionsorgane* 1, Berlin, Springer-Verlag, 1926, pp. 602-618.

Hornbostel, E. M. von, and Wertheimer, M., Über die Wahrnehmung der Schallrichtung, *Akademie der Wissenschaften, Berlin, Sitzungsberichte*, 1920, 15, 388-396.

Janovsky, W., and Spandöck, F., Aufbau und Untersuchungen eines schallgedämpften Raumes, *Akustische Zeitschrift*, 1937, 2, 322.

Jung, Richard (1) Microphysiology of the cortical neurons and its significance for psychophysiology, *Anales de la Facultad Medicina*, Montevideo, 1959, 44, 323-332. (2) Korrelationen von Neuronentätigkeit und Sehen, *in* Richard Jung and Hans Kornhuber, eds., *Neurophysiologie und Psychophysik des visuellen Systems*, Berlin, Springer-Verlag, 1961, pp. 410-435.

Katsuki, Y., (1) Electrical responses of auditory neurons in cat to sound stimulation, *Journal of Neurophysiology*, 1958, 21, 569-588. (2) Pitch discrimination in the higher level of the brain, *International Audiology*, 1962, 1, 53-61.

Katz, David, Methoden der Untersuchung des Vibrationssinnes, *in* E. Abderhalden, ed., *Handbuch der biologischen Arbeitsmethoden*, Abt. 5, Teil, 7, Berlin, Urban & Schwarzenberg, 1937, pp. 879-918.

Keidel, W. D., (1) Messung der Hautwellengeschwindigkeiten bei Vibrationsreizen am Menschen, *Pflügers Archiv für die gesamte Physiologie*, 1952, 255, 213-227. (2) Aktionspotentiale des N. dorsocutaneus bei niederfrequenter Vibration der Froschrückenhaut, *Pflügers Archiv für die gesamte Physiologie*, 1955, 260, 416-436. (3) Synchronization and volley theories, in *Vibrationreception: Der Erschütterungssinn des Menschen*, Erlangen, Universitätsbund, 1956.

Kingsbury, B. A., A direct comparison of the loudness of pure tones, *Physical Review*, 1927, 29, 588-600.

Kohata, Tadashi, Suppression of color contrast and retinal induction by mechanical pressure, *Tohoku Journal of Experimental Medicine*, 1957, 65-66, 239-250.

Kohlrausch, F., Über die Fortpflanzungs-Geschwindigkeit des Reizes in den menschlichen Nerven, *Zeitschrift für rationelle Medicin*, Reihe 3, 1866, 28, 190-204.

Küpfmüller, K., (1) Schwachstromtechnik, *in* W. Wien, and F. Harms, eds., *Handbuch der Experimentalphysik*, vol. 11, part 3, Leipzig, Academische Verlagsgesellschaft m.b.H., 1931. (2) *Die Systemtheorie der elektrischen Nachrichtenübertragung*, Zurich, S. Hirzel Verlag, 1949.

Lamb, Harold, *Hydrodynamics*, Cambridge, University Press, 1932.

Lehmann, Alfred, *Die körperlichen Äusserungen psychischer Zustände*, Teil 3, *Elemente der Psychodynamik*, trans. by F. Bendixen, Leipzig, O. R. Reisland, 1905.

Lissmann, H. W., and Machin, K. E., The mechanism of object location in *Gymnarchus niloticus* and similar fish, *Journal of Experimental Biology*, 1958, 35, 451-486.

Mach, Ernst, (1) Über die Wirkung der räumlichen Vertheilung des Lichtreizes auf die Netzhaut, I, *Akademie der Wissenschaften in Wien, Sitzungsberichte, math.-nat. Cl.*, 1865, 52, Abt. 2, 303-322. (2) Über die physiologische Wirkung räumlich vertheilter Lichtreize, *ibid.*, 1866, 54, Abt. 2, 393-408, and 1868, 57, Abt. 2, 11-19.

Marmont, G., and Wiersma, C.A.G., On the mechanism of inhibition and excitation of crayfish muscle, *Journal of Physiology*, London, 1938, 93, 173-193.

Meissner, Georg, Untersuchungen über den Tastsinn, *Zeitschrift für rationelle Medicin*, 1859, 7, 92-118, esp. 99.

Mountcastle, V. B., and Powell, T.P.S., Neural mechanisms subserving cutaneous sensibility, with special reference to the role of afferent inhibition in sensory perception and discrimination, *Bulletin of the Johns Hopkins Hospital*, 1959, 105, 201-232.

Müller-Lyer, F. C., Optische Urtheilstäuschungen, *Archiv für Physiologie*, 1889, Suppl.-Band, 263-270, and Table 9.

Muralt, A. von, *Neue Ergebnisse der Nervenphysiologie*, Berlin, Springer-Verlag, 1958.

Pfaffmann, Carl, Gustatory afferent impulses, *Journal of Cellular and Comparative Physiology*, 1941, 17, 243-258.

Pohl, R. W., *Mechanik und Akustik*; *Einführung in die Physik*, Springer-Verlag, Berlin, 1932, English trans. *Physical Principles of Mechanics and Acoustics*, London, Blackie & Son.

Polyak, S. L., *The Retina*, Chicago, University of Chicago Press, 1941.

Ratliff, Floyd, *Mach Bands: Quantitative Studies on Neural Networks in the Retina*, San Francisco, Holden-Day, 1965.

Ratliff, Floyd, and Hartline, H. K., The response of limulus optic nerve fibers to patterns of illumination on the receptor mosaic, *Journal of General Physiology*, 1959, 42, 1241-1255.

Ratliff, F., Hartline, H. K., and Miller, W. H., Spatial and temporal aspects of retinal inhibitory interaction, *Journal of the Optical Society of America*, 1963, 53, 110-120.

Rayleigh, Lord, *The Theory of Sound*, New York, Dover, 1945.

Reichardt, W., and MacGinitie, Gordon, Zur Theorie der lateralen Inhibition, in *Kybernetik*, 1962, 1, 155-165, esp. p. 163, Fig. 5.

Renshaw, B., Influence of discharge of motoneurons upon excitation of neighboring motoneurons, *Journal of Neurophysiology*, 1941, 4, 167-183.

Révész, Géza, System der optischen und haptischen Raumtäuschungen, *Zeitschrift für Psychologie*, 1934, 131-132, 296-375.

Riggs, L., Ratliff, F., Cornsweet, J. C., and Cornsweet, T. N., The disappearance of steadily fixated visual test objects, *Journal of the Optical Society of America*, 1953, 43, 495-501.

Riggs, L. A., Ratliff, F., and Keesey, U. T., Appearance of Mach bands with a motionless retinal image, *Journal of the Optical Society of America*, 1961, 51, 702-703.

Rose, J. E., and Mountcastle, V. B., Touch and kinesthesis, *in* John Field, ed., *Handbook of Physiology*, Section 1, *Neurophysiology*, 1, Washington, D.C., American Physiological Society, 1959, pp. 387-426, esp. p. 409, Fig. 12.

Ruch, Theodore C., Somatic sensations, *in* T. C. Ruch and J. F. Fulton, eds., *Medical Physiology and Biophysics*, Philadelphia and London, W. B. Saunders Co., 1960, pp. 300-322.

Savart, F., Über die Ursachen der Tonhöhe, *Annalen der Physik und Chemie*, 1840, 51, 555-561.

Sechenov, I., *Selected Works*, Moscow-Leningrad, State Publishing House for Biological and Medical Literature, 1935.

Sherrington, Charles, *The Integrative Action of the Nervous System*, New Haven, Conn., Yale University Press, 1947.

Stern, W. L., Psychische Präsenzzeit, *Zeitschrift für Psychologie*, 1897, 13, 325-349.

Stevens, Stanley S., ed., *Handbook of Experimental Psychology*, New York, John Wiley and Sons, 1951.

Suga, N., Analysis of frequency-modulated sounds by auditory neurones of echo-locating bats, *Journal of Physiology*, 1965, 179, 26-53, esp. p. 30, Fig. 2.

Taylor, G. W., The optical properties of the shrimp nerve fiber sheath, *Journal of Cellular and Comparative Physiology*, 1941, 18, 233-242.

Tunturi, Archie R., A difference in the representation of auditory signals for the left and right ears in the iso-frequency contours of the right middle ectosylvian auditory cortex of the dog, *American Journal of Physiology*, 1952, 168, 712-727.

Vierordt, K., *Der Zeitsinn nach Versuchen*, Tübingen, H. Laupp, 1868.

Weber, Eduard F. W., and Weber, Ernst H., Experimenta physiologica in theatro anatomico, *Annali Universali di Medicina, Milano*, 1845, 116, 225-233, esp. Experimenta, quibus probatur nervos vagos rotatione machinae galvano-magneticae irritatos, motum cordis retardare et adeo intercipere, pp. 227-228.

Wiener, Otto, Darstellung gekrümmter Lichstrahlen und Verwerthung derselben zur Untersuchung von Diffusion und Wärmeleitung, *Annalen der Physik und Chemie*, New Series, 1893, 49, 105-149.

INDEX